Engineering for Industrial Designers and Inventors
Fundamentals for Designers of Wonderful Things

Thomas Ask

Beijing · Boston · Farnham · Sebastopol · Tokyo

Engineering for Industrial Designers and Inventors

by Thomas Ask

Copyright © 2016 Thomas Ask. All rights reserved.

Printed in the United States of America.

Published by O'Reilly Media, Inc., 1005 Gravenstein Highway North, Sebastopol, CA 95472.

O'Reilly books may be purchased for educational, business, or sales promotional use. Online editions are also available for most titles (*http://safaribooksonline.com*). For more information, contact our corporate/institutional sales department: 800-998-9938 or *corporate@oreilly.com*.

Editors: Nan Barber and Susan Conant
Acquisitions Editor: Laurel Ruma
Production Editor: Shiny Kalapurakkel
Copyeditor: Kim Cofer
Proofreader: Jasmine Kwityn

Indexer: Angela Howard
Interior Designer: David Futato
Cover Designer: Karen Montgomery
Illustrator: Rebecca Demarest

April 2016: First Edition

Revision History for the First Edition
2016-04-27: First Release

See *http://oreilly.com/catalog/errata.csp?isbn=9781491932612* for release details.

The O'Reilly logo is a registered trademark of O'Reilly Media, Inc. *Engineering for Industrial Designers and Inventors*, the cover image, and related trade dress are trademarks of O'Reilly Media, Inc.

While the publisher and the author(s) have used good faith efforts to ensure that the information and instructions contained in this work are accurate, the publisher and the author(s) disclaim all responsibility for errors or omissions, including without limitation responsibility for damages resulting from the use of or reliance on this work. Use of the information and instructions contained in this work is at your own risk. If any code samples or other technology this work contains or describes is subject to open source licenses or the intellectual property rights of others, it is your responsibility to ensure that your use thereof complies with such licenses and/or rights.

978-1-491-93261-2

[LSI]

Table of Contents

Preface ix

1. Design! 1
   Intuition   1
   Designing for People   3
   Design Process   3
   Design by Building   4
   The "Design Thinking" Process   5
   Battle with the Virtual World   7
      The Blank Sheet of Paper   8
   Starting the Design   9
   Philosophical Foundations   10

2. Nontechnical Influences. 13
   The Role of Aesthetics in Design   13
   The Need for Brand Management   19
   Material Culture: The Context of Design   21
   The Influence of Tradition   22
   Problems with Tradition   22
   Visual Stereotype: Familiarity Influencing Design   22
   Ethnography's Role in User-Centered Design   24
   Creativity: The Fount of Design   29
   What Is Creativity?   31

3. Material Mechanics. 33
   The Effects of Pulling, Pushing, and Twisting Forces   34
   Static Loads   36
   Mechanics of Deformable Bodies   39
   Shape   43

      Relationship Between Shape, Stiffness, and Flexural Stress   43
      Relationship Between Stress and Thickness   44
   Stress Concentration   47
   Fatigue   49
      Fatigue Strength   50
      Fretting Fatigue   51
      Creep and Thermal Relaxation   51
   Buckling   52
   Thermal Expansion   54
   Failure Modes   57
      Elastic Deformation   57
      Yielding   57
      Ductile Rupture   57
      Brittle Fracture   58
      Fatigue   58
      Impact   58
      Buckling   59
      Thermal Shock   59
      Wear   59
      Brinelling   59
      Spalling   60
   Noise and Vibration   61
      Noise   61
      Vibration   62
   Closing Thoughts   62

4. Materials   65
   Characteristics of Metals   67
      Steel   68
      Common Steel Alloys   69
      Stainless Steel   69
      Aluminum   70
      Bronze   72
      Nickel   72
      Magnesium   72
      Titanium   73
      Zinc   73
   Characteristics of Ceramics   75
   Characteristics of Plastics   75
   Characteristics of Foams   76
   Characteristics of Wood   77
   Characteristics of Composite Materials   78
      Characteristics of Fibers   79
   Chacteristics of Adhesives   80
   Corrosion Behavior of Materials   82

Anodes and Cathodes	82
Rust	83
Crevice Corrosion and Pitting	85
Intergranular Corrosion	85
Stress Corrosion Cracking	85
Biological Corrosion	85
Corrosion Characteristics	86
Closing Thoughts	88

5. Thermodynamics ... 89
The Baseline Temperature	90
Pressure	90
Ideal Gas Law	91
Thermodynamic Laws	91
First Law of Thermodynamics	91
Second Law of Thermodynamics	95
Third Law of Thermodynamics	95
Zeroth Law of Thermodynamics	95
Nature of Phases	95
Combustion	103
Fuels	105
Closing Thoughts	108

6. Fluid Mechanics ... 109
Fluid Behavior	109
Fluid Statics	110
Pneumatics and Hydraulics	111
Fluid Dynamics	112
Rheology	112
Boundary Layer	114
Drag	116
Closing Thoughts	120

7. Heat Transfer ... 121
What Is Heat?	121
Conduction	122
Convection	122
Radiation	123
Nature of Electromagnetic Radiation	123
Radiation Heat Transfer	124
Combined Effects with Radiation	128
Radiation, Conduction, and Specific Heat	128
Radiation and Convection	129
Other Heat Transfer Considerations	132
Mass Transfer	132

	Cause of Infiltration	133
	Thermal Storage	134
	Internal Sources of Heat	135
	Closing Thoughts	138

8. Human Factors... 139
Ergonomics 139
Handle Design 140
Anthropometry 140
Designing with Anthropometric Data 142
   Seat Design Example 143
Kinesiology 144
   Walking and Running 144
Mapping 145
Human Injuries Related to Design 146
Biomimicry 148
Interaction Design 148
Expert Systems and Artificial Neural Networks 149
Closing Thoughts 150

9. Sustainability in Design.............................. 151
Backstory 152
Usage 154
Disposal 154
Closing Thoughts 157

10. Mechanical Systems.................................. 159
Pumps, Compressors, and Fans 160
   Positive Displacement 160
   Dynamic 161
Electric Motors 162
Transferring Power 163
Pneumatics and Hydraulics 163
   Pneumatics 164
   Hydraulics 164
   Actuators 165
Gears 165
Belts and Chains 168
   Belts 169
   Chains 170
Cable Drive 170
Brakes, Clutches and Shaft Drives 171
   Brakes and Clutches 171
   Shaft Drives 173
Closing Thoughts 173

11. Mechanical Components....................................	175
Bearings	175
Gaskets and Seals	177
Fasteners and Bonds	179
Threaded Fasteners	179
Rivets	182
Shaft and Terminal Fasteners	183
Bonded Joints	183
Welds	183
Interference Fits	184
Springs	185
Closing Thoughts	186
Epilogue..	187
Index..	191

Preface

We are all impatient. We want everything at our fingertips on demand. What this book does is cut to the core elements of mechanical engineering without interrupting your study with too much noise. The theory is rigorous, the math is not. I prefer qualitative descriptions and analogies over math. The goal of this book is to make mechanical engineering a real tool in your inventive, creative tool belt. I want you to understand basic concepts that can inform the copious decisions required when designing a real object. Reading a book is certainly a test of patience, and I admire your foray into this walk together. Your time is under assault by many things. Your slices of open time are presented with enticing tweets and games so that your time is not linear and logical but rather a polymodal blend of grand endeavors and tiny treats. Where is the time for reflection?

Death by Video

Drunken eyes seducing minds
Flashing fluorescence digging

Synapses fire on and off
on and off
on and off

Eyes blink only for water
Gently cooling heated sight
Washing across digital rough
As the synapses fire on and off

Chant a dirge or scumble color
While watching forced slumber

Noises zipping, busy, fuzzy dots
Blinking, blanching, watching plots

Jam the mind in a keyboard crack
Waiting on your electro-sausage.

Synapses. fire. on. and. off.
on. and. off.
on. and. off.

—Thomas Ask (Excelsior Review, Issue No. 5, April 2015)

This book is for creators, content providers, inventors, and those who create mechanical marvels from the void. It is for those who work in all three dimensions and strive to take their wonderful ideas and move them to reality. We all wish to be experts in something. We self-identify our qualities, develop a group affiliation, and protect that in which we take pride. The often unspoken secret is that we can be good in multiple things—we can learn, adapt, and morph as we fight complacency.

Many of us bounce ideas around in our mind and with friends and family. These ideas bring us great promise, joy, and wonder. However, turning ideas into functional products requires specific technical knowledge. During the creative process of design, this knowledge needs to be visceral and drawn from the fingertips like yellow, crackling lightning bolts. You don't want to have to look up information and analyze something, you just want to design. You want to stand on the shoulders of giants, mixing your delightful creativity with engineering.

Why I Wrote This Book

I wanted to open up the world of mechanical engineering to others. I saw many people with great ideas who couldn't work out the engineering polish on their designs. They would design a feature that was likely to break or distort. They would make some feature that would discourage cooling or cause high drag.

My mechanical engineering education was a theoretical, calculus-based presentation of concepts. Much time was spent in deriving equations and quantitative analyses. When I worked as a designer, I found that this knowledge didn't serve me well. I couldn't analyze something I hadn't designed. Even after concepts are agreed upon, the detailed design of a part requires many decisions that allow it to be lighter, stronger, cooler, or whatever the design goals are. You had to put your best guess up for analysis. The multiple design decisions on an object's physical qualities (thickness, radius, roughness, etc.) are based on having an intuitive feel for how solids, liquids, and gases behave.

Typically, design methodology calls for a sequence of events, which in their simplest form are the clear definition of design requirements, development of concepts, and finally engineering the final design and manufacturing approach. However, many more ambiguous factors get introduced into the design process. Because design is done by humans and typically for human use, a long list of factors are introduced into the development of designs, ranging from the sociological forces of group identity and organizational behavior to an organization's politics, individual egos, and ethics. Aesthetics assert a subliminal force even on the most nonconsumer-oriented design. Moreover, technical and business issues arise to provide design direction, such as ergonomics, performance, longevity, capital costs, and profitability. Regulatory forces such as safety standards often provide some baseline for starting a design while corporate attitudes toward environmental sustainability and manufacturing preferences can suggest design approaches.

Traditionally, design requires technical skills and experience to ensure that the product or system works in the way it is intended. Therefore, product design is often left to the engineers. However, intangible forces become integrated into the design and direct the final appearance of the design. The sometimes subliminal forces can be brought to light through an interdisciplinary study such as occurs in the industrial design discipline. Although industrial design has technical aspects, the discipline carefully considers aesthetic, social, cultural, and organizational forces. This book should ground industrial designers and inventors in mechanical engineering but it should open up other lines of inquiry—the nontechnical affairs that make a product fun and relevant. I think all good designs have an element of surprise and playfulness. We all have the capability to engage many disciplines in pursuit of excellence in creating products.

How This Book Is Organized

Designers consider every single aspect of a design and the user experience. Nothing is left out—from the sound and feel to the particular requirements determined by anthropometry and kinesiology. Machine design is result-oriented rather than human-oriented. However, a machine designer can certainly have a wonderful sense of aesthetics and ergonomics, while an industrial designer can have a powerful ability to understand complex systems and a strong conceptual grasp of engineering principles.

The book starts with an overview of design and nontechnical aspects. It then moves into traditional mechanical engineering topics and some related considerations. Finally, the book offers a flavor for the parts and pieces that go into mechanical devices.

Specifically, Chapters 1 and 2 provide a context for mechanical design. They address the human-centric, nontechnical aspects of design. These are important ideas, ones that should be intuitive to some degree. However, we also consider aesthetics, creativity, brand management, material culture, tradition, visual stereotypes, ethnography, and design processes.

Chapters 3 through 7 consider the traditional mechanical engineering disciplines related to the behavior of solids, liquids, and gases. Chapter 8 considers ergonomics, which is surprisingly easy to embed in designs but is very important. Chapter 9 puts a bookend on the theory portion as we consider sustainability.

Chapters 10 and 11 introduce real, practical systems and components. These chapters do not strive to compete with the excellent design guides offered by vendors but they do offer an overview of critical parts and pieces that can make a design successful.

A Word on Nomenclature

The word *design* can mean a lot of things, but in this book it is used to indicate the nebulous process of creating something mechanical, whether it be traditional industrial/product design or inventing. This activity differs from engineering analyses or any codified approach to developing a product. While electronics resides in almost all components, it is not addressed here. However, inventive types need to be familiar with control systems and electronic componentry.

A Word on Jargon

A good way to disenfranchise someone is to use a different language. We can all tap into a reservoir of words that assert our knowledge of something. The words can be dragged out from academic disciplines, company jargon, or hipster idioms. I have tried to use rich, discipline-specific language so if you are not a member of the club already you can join in. While some of the terminology can be off-putting, I didn't make the words up. The technical language provides entrée into these engineering communities.

Who Should Read This Book?

This book is for inventive people who wish to integrate engineering into their creative design process. Engineering concepts are presented in a rigorous but largely nonmathematical format. Mind stories, images, and key concepts are developed that help a reader successfully weave engineering principles into the design process. This book can help anyone move sketches to engineered products.

Specifically, this book is intended for students and practitioners of product and mechanical design as well as the technically minded, inventive public.

Safari Books Online

Safari Books Online is an on-demand digital library that delivers expert content in both book and video form from the world's leading authors in technology and business.

Technology professionals, software developers, web designers, and business and creative professionals use Safari Books Online as their primary resource for research, problem solving, learning, and certification training.

Safari Books Online offers a range of plans and pricing for enterprise, government, education, and individuals.

Members have access to thousands of books, training videos, and prepublication manuscripts in one fully searchable database from publishers like O'Reilly Media, Prentice Hall Professional, Addison-Wesley Professional, Microsoft Press, Sams, Que, Peachpit Press, Focal Press, Cisco Press, John Wiley & Sons, Syngress, Morgan Kaufmann, IBM Redbooks, Packt, Adobe Press, FT Press, Apress, Manning, New Riders, McGraw-Hill, Jones & Bartlett, Course Technology, and hundreds more. For more information about Safari Books Online, please visit us online.

How to Contact Us

Please address comments and questions concerning this book to the publisher:

O'Reilly Media, Inc.
1005 Gravenstein Highway North
Sebastopol, CA 95472
800-998-9938 (in the United States or Canada)
707-829-0515 (international or local)
707-829-0104 (fax)

We have a web page for this book, where we list errata, examples, and any additional information. You can access this page at *http://shop.oreilly.com/product/0636920042617.do*.

To comment or ask technical questions about this book, send email to *bookquestions@oreilly.com*.

For more information about our books, courses, conferences, and news, see our website at *http://www.oreilly.com*.

Find us on Facebook: *http:///facebook.com/oreilly*
Follow us on Twitter: *http://twitter.com/oreillymedia*
Watch us on YouTube: *http://www.youtube.com/oreillymedia*

Acknowledgments

Many people have worked hard and contributed their talents into making this book as good as can be. From the style of presentation to the accuracy of information, I have been helped by friends and family.

I wish to thank Anne Reiner, a gifted writer, for her style guidance at the early stages of this writing. I also wish to thank my reviewers who shared their deep knowledge and precious time in reviewing and correcting my work. I am grateful to these entrusted colleagues: Tim Brauning, Henry Kang, and Georgene Rada. I also wish to thank my O'Reilly editor, Nan Barber, who offered a calm voice and professional guidance throughout the writing process.

All my love and gratitude goes to my wife, Beth, who has always been my counsel for all things and was again asked to review this work. My kids, Eric and Elayna, taught me a lot about applying engineering and design through our play and experimentation. Thank you all!

CHAPTER 1

Design!

I HAVE DESIGNED DOZENS OF COMMERCIALIZED PRODUCTS AND SYSTEMS DURING MY CAREER, and none have started with an equation. They more often start with a walk. This initial reflection is followed by sketches, conversation, and a simple model until finally a design concept is developed. Upon this ascendant design, the rich technical apparatus of engineering and computers can be brought to bear.

Designers create order out of chaos, but design is not science—we need more than science in our quest to manipulate our environment. The goal of this book is to develop the "gut feel" and robust conceptual theory that can be drawn into the design process. The creative process of design is moderated by the physical and life sciences, but you can't do mathematical modeling and computer simulation during brainstorming sessions.

Designers also appeal to more than the technical and mechanistic. We recognize that people surround themselves with beauty, whether the sparkle of a diamond or a poster of the Eiffel Tower. Functional sculpture can arise from the gifted and informed hands of the designer, but designing without a grasp of engineering fundamentals is like digging a hole without a shovel.

Intuition

The notion of "seeing air movement" or "feeling how a material behaves" is not a sophomoric approach to design—it is a necessary one. Design concepts can be proven or disproven by intuitive feel or simple experimentation. For example, when designing a mechanical linkage, there is no dishonor in cutting out linkages from foam and using nails as hinges to produce a model. It is fast, intuitive, and complements ideation. I have presented a modified kid's toy to a client to illustrate a new idea and conducted tests in my driveway or garage with all sorts of assembled widgets before showing them to the boss (wow, it worked the first time I tried it at the office!).

The designs of the SR 71 Blackbird and other famous planes were headed by Kelly Johnson of Lockheed Martin Skunk Works. He designed by "seeing air" and using seat-of-the-pants approaches. I remember my first day as an 18-year-old intern where I was asked to improve the efficiency of an engine heat exchanger. I didn't know anything about fluid dynamics or heat transfer equations but I did have a sense for how air flowed, which was all I needed to create the new design. The engineering analyses followed the design concept. Later in my career, I designed an engine starting system for use at arctic temperatures. I designed a high-torque air motor, used lots of fancy lubricants, and incorporated off-the-shelf technology like ether injection to come up with a system that worked. It was not a single product that made this work; it was a combination of product designs, systems, and control systems.

Helicopter blade material selections illustrate the need for qualitative approaches in design. As shown in Figure 1-1, they are comprised of a variety of materials. The blade is typically made by wrapping an aluminum D spar and honeycomb core in fiberglass or carbon fiber of different orientations. Upon this structure, a chromium steel erosion shield is adhered. In addition, blades can also include balancing tubes, lightning receptors, and a variety of means for inflight de-icing. Some of these materials make the blade light, some make it strong, and some contend with bird and lightning strikes. None of these amalgams would be specified in a textbook.

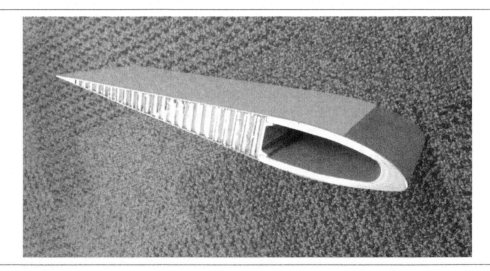

FIGURE 1-1. Blade cross section

The art of design requires us to use engineering where it is helpful and rely on the social sciences where they can guide interpretation. Design, therefore, does not derive from some big math equation or a focus group's impractical concept. We bounce between the humanism of Cicero and the Ancients to the rationalism of the Enlightenment, which set science on the throne. Engineering has a positivistic framework where the scientific method prevails. We should recognize that design is founded on more interpretive epistemologies. We need to recognize our relationship to information so that we try to reduce the influences of our personal experiences and cultural biases from our interpretation of the information.

Design draws upon the humanist arguments such as Protagoras' assertion that "man is the measure of all things," and Edmund Husserl's understanding of scientific method's limitations and its inappropriateness for assessing human thought and actions. Besides being a philosopher, Blaise Pascal was a gifted mathematician and physicist. He opined, "The heart has its reasons, which reason does not know." That is, we grasp truth beyond our reasoning ability. We can't decipher the reason for the allure of a poem or work of art; they stand apart from the world of science. Art exists in design and this aspect of design is difficult to approach in a structured, positivist manner.

Designing for People

Recently, I designed a fishing boat for a Southeast Asian application based on ethnographic approaches. This endeavor was an example of industrial design where the functional and nonfunctional (mechanistic and nonmechanistic) need to blend as a coherent whole. For example, I had to recognize that fishermen trusted wood because it floats, and in the event of a sinking, this wood flotsam created their only life raft. These boats lasted "forever" and fishermen had no great interest in fashion statements. However, the boat had to do more than function—it had to appeal in many subtle ways. Appeal doesn't mean pretty. Many products are decidedly uncute but are still appealing. While aesthetics need to be compelling, the material culture and traditions need to be incorporated into designs.

We manage our lives efficiently within our personal world view. We stitch experience and knowledge to make sense of the huge amount of information we receive. I was following a guided tour at the Cooper Hewitt museum recently. The guide was young. In the audience was a well-dressed lady holding a clipboard. I put this data together to surmise that the guide was being assessed by her employer. If the guide were older and I saw a young person in the audience, informally dressed and holding a spiral notebook, I would have a different conclusion. You may note that a person wearing a hard hat and safety vest (perhaps coupled with a white van) can go anywhere and do anything.

Pursuing nontechnical aspects of design requires ethnography, the scientific study of people. Ethnography is a powerful tool for all designers of wonderful things, and this is addressed in Chapter 2.

Design Process

Declaring a process by which things are designed is impossible. Many organizations have carefully considered processes they find very helpful and visualization tools exist to support idea creation (ideation). However, these approaches and tools change with time. In contrast to the ephemeral, idealized design process and tools, a helpful theoretical construct for the design process lies in part with the notion of bounded rationality. This concept asserts decisions need to be made with partial, sometimes fragmentary information, where feedback is achieved only in the future. Therefore, value has to be attributed to the decisions immediately. This theory further recognizes that not all alternatives are considered before a decision is proffered. Heidegger's form of the dialectic in which you continually re-ask the same question is consistent with the argument that in contending with the concept of bounded rationality you must replace the optimum with the sufficient. However, engineering theory offers an immutable foundation for the bounded rationality required in design.

The creative design process requires rapid distillation of numerous ideas and this process can fail to consider good design concepts. Most notably, a designer is required to use good judgment and be confident in the results. For most businesses, it is impractical to develop a thousand design ideas, convene a thousand focus groups, and proceed with the "best" design that derives from this expensive and time-consuming process. Rather, the designer is encumbered by the fact that the target user does not necessarily know what he or she specifically wants, even though they are usually sure as to what they do not want. In this respect, design is not fully rational and the notion of bounded rationality provides a good construct for understanding how ambiguity must be tolerated in the design process.

Design by Building

Socrates opined that a life not examined is not worth living. The importance of reflection is highlighted by the educational theorist David Kolb's description of experiential learning. The Kolb model presents an amalgam of perception and processing. This model suggests that deep learning derives from moving through four waypoints: feeling, watching, thinking, and doing. Cycling through these waypoints, where the doing informs the thinking, highlights the benefit of experiential learning, especially when personal reflection arises during the course of the learning process. Experimenting and making mistakes allows you to develop tacit knowledge—knowledge that might be captured in "rules of thumb" or "mistakes made." This book only provides part of the mechanical arts. You need to leave the textbook page and CAD program, then build stuff, and make mistakes from which you can learn.

Buckminster Fuller's daughter, Allegra Fuller Snyder, said that her father based everything on experience. "I believe inherently Bucky's concept of mind has, at its base, mind processing through experience... his fingertips were his antennas to experience. No idea that he processed in his mind was ever processed without that link to experience." Designing by building can have various outcomes, from the successes of Fuller to the lesser success of the Bent Pyramid of Snefru (Figure 1-2), in which case the builders recognized they couldn't maintain such a steep angle for the pyramid as they originally envisioned.

FIGURE 1-2. Bent Pyramid of Snefru

While designers are striving for some goal, they must often allow the changes motivated by the physical building process. This playfulness with design allows the visceral to mix with the abstract, the hands and mind working in concert to achieve something unplanned. When Leonardo applied his fifth glaze of paint on the *Mona Lisa*, did he anticipate the results of his twentieth glaze? Did Fuller realize how his cut and bent paper would end up as a dome? Building is a design tool with a human hand and eye.

The "Design Thinking" Process

"Design thinking" is a method for addressing complex problems that are difficult to define. These can be a product (e.g., a nail clipper for an 8-year-old Norwegian girl) or systemic (e.g., how to educate high school students). Design thinking promotes the rapid analysis and synthesis of ideas. This process strives to find insights and patterns into high order, poorly defined problems with obscure relationships.

The procedure cycles through a series of steps: (1) empathy, (2) problem definition, (3) ideation, (4) prototyping, and (5) testing.

In other words:

1. Learn about the users and other stakeholders
2. Gain user insight, their point of view
3. Brainstorm ideas
4. Take the best ideas and make them tangible by building models
5. Actually try things out, don't just talk about them

This process draws upon ethnographic techniques to gain insights into the stakeholder's point of view. This process recognizes the social nature of design—we want to either use a design with other people or display it publicly. I have conducted design thinking practice sessions in which the ethnographic component is simply interviewing a person who has unique knowledge about something. For example, we have designed a storage bag for a specific car model and a food dish for a Hari Raya celebration. We cycle through the process of learning about the stakeholders' point of view, sketch ideas, and then use fabric, paper, tape, and markers to prototype something.

In practice sessions, we normally depict the problem as a sequence of drawings, called a storyboard, as shown in Figure 1-3. This is like a comic strip that shows how something is used step by step. We also use experience maps that are more freeform but capture the time-related experience with some object or process. We then try to identify specific pain points in the experience and agree with the interviewee on the root problem. Next, we develop sketches that address the agreed-upon problem. Finally, we draw a new experience map or storyboard with the improved idea prior to prototyping and testing the idea. Most problems don't have a single stakeholder and broader ethnographic techniques are required. Moreover, the feedback loop is more cumbersome than dealing with a single representative user.

FIGURE 1-3. An example of a storyboard (image courtesy of Elizabeth Snedeker)

Battle with the Virtual World

I struggle to unglue myself from the virtual world. Recently, I wanted to make a didgeridoo after being impressed by both the sight and sound of this bamboo-based instrument at a local music store. This looks pretty simple, I thought. I could take some three-inch PVC pipe and quickly create this instrument. I was off with my kids to my workshop where I could have fun making this thing. The joy of discovery and the disappointment of failure have always been mixed for me during these endeavors. But before starting this project, I first turned on my computer, which seems to lower my blood pressure for some inexplicable reason. After doing the compulsory email, news, and weather checks, I went to YouTube. With a few key clicks, I looked up didgeridoos. The Internet was awash in my idea of making didgeridoos. I guess I wasn't the first one to think of a homemade didgeridoo made out of pipe. I watched a couple of "how to" videos. Then I became visually saturated with didgeridoo making and moved on to another part of my cyber world quest, the notion of actually making a didgeridoo abandoned. I had seen how they were made by others, how they sounded, and was placated by this visceral journey—typing is so much easier than actually working with pipe, torches, and tools.

    The joy of discovery, the venturing into the unknown—this is what I like. Socrates did too, saying "wisdom begins in wonder." Neil Armstrong agreed, saying, "mystery creates wonder and wonder is the basis of man's desire to understand." Yet the Internet makes discovery and understanding so easy that now I would rather type in keywords than actually hunt down my quarry. It seems sad that discovery has become passé. Why not Google it and move on?

Can artificial intelligence kill aesthetic exploration, where suggested videos, music, and the like are accepted as wiser than serendipitous wanderings and bold experimentations? In like manner, embedded engineering analysis in parametric modeling and default fillet radii seem to take more and more from human decision making. We start trusting algorithms and they start guiding us, thereby reinforcing their effectiveness.

The virtual world has invaded other areas of fun, adventure, and exploration. I recently hosted a kite-propelled boat race in which students were asked to build a boat from scratch that could carry one person and be propelled by a kite. In the face of computer games, this activity didn't have a chance. So my workshop, with refreshments awaiting and fun to be had, was empty except for three diehards, who left quickly and quietly to the world of the virtual where fantastic kingdoms and raging videos are finger movements away. Why build a floating hunk of garbage that probably won't look good and perform even worse? Of course the answer is that it satiates a human need to explore, build, and trust that which we built. To make mistakes, cut fingers, and flop down into a frayed chair in frustration from a failed idea. To work through problems and deal with things we can touch and drop on our foot. Someone has to make content for the virtual world. Who is left?

However, all is not so glum. I had eight students participate in our kite-driven boat race and almost everyone got joyously wet and cold.

THE BLANK SHEET OF PAPER

Every professional designer's joy and fear is a blank sheet of paper.

Ideation in a classroom studio is good fun as your mind gets to fly about and a surging rush of creation can be felt in the throat. However, in professional practice, you feel this surging rush in your gut also. You realize how arbitrary the whole process is. A small, whimsical thought can lead a team down a hallway of design that cannot easily be reversed. You worry about what you have not thought about. Is yours the "great idea" or are you going to miss your mark and come up with something stupid? You realize that the initial surge of ideas will lead to a concept that many people will spend a lot of time working on.

One of my most boring designs (a natural gas fitting) has been sold for over 20 years and is still out there. I remember looking at the blank sheet of paper on my drawing board and thinking, *now what?* I remember one group of designers I convened to redesign a product and we could not for the life of us think of a better idea than what was already out there. I felt humbled and thought that I was missing some great insight. In contrast, one of my favorite brainstorming experiences was a night walk on a beach with a couple of colleagues where we generated a flow of great, practical inventions. Practical ideas that address a specific need and lead to commercial products are far different than blue sky designs of beautiful chairs and jazzy automobiles.

So, how do you start? How do you deal with a blank sheet of paper? Some discoveries are accidents and we should be appreciative of the burst of creation. No prescribed path was taken nor can a backstory be described.

Thou that hast giv'n so much to me,
Give one thing more, a gratefull heart:
See how Thy beggar works in Thee
By art:

—George Herbert (Gratefulnesse, 1633)

Starting the Design

The notion of paper is not an antiquated, quaint homage to Leonardo da Vinci. Even experienced CAD operators who have grown up with this medium still prefer the humble paper and pencil.

A design can start with what you know. This might include the following:

1. Industry standards that may apply
2. Required dimensions (to fit people or other equipment)

While these approaches can be a noose that constrains creative exploration, if your design is for an aftermarket product that must attach to a bicycle handlebar, you must first find out the dimensions of existing handlebars. If your design is to be used by humans, you need to identify suitable dimensions. The study of human dimensions, called anthropometry, will be presented in Chapter 8.

Other established information can also help design. The design must be mechanically sufficient without holes and notches where the highest stresses might develop. Understanding the typical shape of the object people are familiar with seeing is also helpful. This *visual stereotype* concept will be described further in Chapter 2. Visual stereotypes should be the least constraining metric, but this concept helps you understand what people expect to see in a design. The visual stereotype can provide a helpful starting point, especially in well-established markets. However, surprise is one mark of creative design so breaking a visual stereotype can often be wise.

The human appeal of design elements such as symmetry, repetition, rhythm, balance, proportion, harmony, movement, color, and texture, are tools in your arsenal—just as color theory can aid in developing color schemes or storyboarding can force you to think through processes. You may find it helpful to think about the numerical aspects of these, such as in proportion, repetition, and composition. Consider experimenting with the "golden ratio" of 1.618 or repetitive, numerically linked ratios. The golden ratio occurs frequently in nature; for example, this is the approximate ratio of the finger bones. The golden ratio appears often and is defined as lengths or areas sized such that the ratio of short length over the long length is equal to the ratio of the long length to the sum of the two lengths. Figure 1-4 shows the two-dimensional relationship of the golden ratio in the Fibonacci spiral.

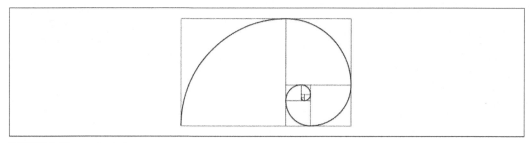

FIGURE 1-4. Fibonacci spiral

Fortunately, there are no specific rules for getting started and addressing the challenge of a blank sheet of paper. Design is delightfully human and you will develop your own personal methods as you experiment with various established methods of ideation and design development.

Philosophical Foundations

Design is an applied science. Reflections on the capability and limitations of science are helpful for understanding potential benefits. We can recognize that not all technical advances are orderly and founded on historical developments. In addition, science can give an incomplete picture of truth. Newton moved beyond the physics of Aristotle and was able to mathematically describe motion in a valid fashion for his context of low-speed, high-mass objects. However, his descriptions of motion were ultimately incomplete. Einstein expanded these descriptions into a "universal" truth by considering relativistic (high-speed, low-mass) effects. Newton's data was valid, relevant, and useful but it was not fully accurate. Designers benefit by recognizing science can be blind to the broad swath of culture or even be trapped within its own mental models. Therefore science fails to provide all data and guidance for culturally appropriate designs.

The professional practice of design is complemented by interpretive approaches that fall outside the natural sciences. Not all questions are scientifically answerable and some can only be judged based on their value to the individual. Immanuel Kant is famous for identifying the distinction between the world as it actually is versus its appearance to us. We look out a window and see a flower, but what we actually see are colors, shades, and shapes. We put this sensory data in the phenomenal realm together to form the idea of a flower. There is a difference between the sensory data and the actual thing.

The two basic divisions of philosophy in connection with the scientific elements of design are associated with Karl Popper and Thomas Kuhn. Kuhn thinks that science is constrained by paradigms and group behavior. Scientists will not abandon a theory based on a single, contrary occurrence (a falsification event). They seek (or await) a new paradigm that accommodates new data; however, they do not do so with alacrity. Popper, on the other hand, asserts that scientists quickly abandon previous theory in the face of evidence. While Popper recognized the limits of scientific exploration, he claimed that science is always testing theories, modifying them, and extending the Socratic method of inquiry by using scientific instruments.

While the challenged relationship between theory and evidence goes back at least to David Hume, who writing in 1748, asserted that no amount of evidence can support a theory because there are an infinite number of predicted outcomes. Philosophers such as Paul Feyerabend have extended this inquiry to the limits of the scientific method. They argue that the outcomes of revolutionary times in science occur when traditional mental models or paradigms are being challenged. These challenges lead to unanticipated consequences that are not guided by science alone. Examples of revolutionary epochs, such as associated with Copernicus and Einstein, arise when science is not funnelled through a paradigm. These profound advances in science are often associated with radical changes in mental models or paradigms. Once these paradigms are established, a period of "normal science" flourishes. However, science will then resist notions that are contrary to the paradigm until it is compelled to do so by escalating evidence.

While philosophers such as Paul Feyerabend and Stansilav Grof would consider the whimsy and chaos of individuals as productive and powerful agents of change, creative speculation does not negate the power of the scientific method, from the practical inventions of Thomas Edison to Richard Feynman's march through theoretical physics. Designers often don't have enough information to make a scientific decision. This "bounded rationality" compels designers to use abductive reasoning with its reliance upon a conclusion (a design that works!) rather than formal rules and preconditions—the ends justify the means. Designers simply throw away bad designs and keep working to get what they want.

Interpretive approaches to knowledge reign outside the "scientific method" type of science described by Popper. Interpretive views expressly note that evidence is not proof of anything, rather the evidence has to be interpreted by someone. Information is routed through the context of cultural and social experiences that are imperfectly parsed by many qualitative analyses methods. These interpretive methods, often derived ethnographically, help answer questions such as: What is strong enough? What is durable enough? What do the stakeholders really want?

All disciplines work in a social context and the intellectual ecology that they operate under motivates behaviors and opinions. We quickly identify the "rules" for our group and try to defend our group identity. While sometimes these rules are written, most are discerned by observing what happens when people break them or by how people actually behave. Violating a group's rules as expressed by stories, traditions, and practices can disturb the intellectual ecology and cause us to be anxious about what we are doing. This anxiety can motivate us to experiment with designs aside like-minded designers or by yourself in your workshop.

Could the desire to find a philosophical opening for design be a waste of time? This overview is intended to highlight the creative impulse to explore the unknown. While the ability to stand on the shoulders of giants through education and shared information has propelled us faster than the Ancients could have imagined, there is a small space for those who wonder and experiment with their quiet whims. Designers are not entrenched in the scientific method, rather they are rooted in human creativity.

Creativity and other nontechnical issues will be discussed in Chapter 2, but as we start looking at the traditional fields of material mechanics, thermodynamics, fluid dynamics, and heat transfer, we need to recognize design looks forward to what has yet to be accomplished. The forward look of design contrasts with the old and reliable analyses that correspond to the forthcoming presentation of engineering disciplines. However, the quantification of these theories are used to analyze that which is already out of the concept stage. These engineering disciplines present concepts that guide designs. They not only look brightly forward to detailed designs, but can also glare intently backward into concept development.

CHAPTER 2

Nontechnical Influences

WHY DOES ONE PEN GET THROWN IN THE GARBAGE AFTER USE, ANOTHER IS WORTHY OF A refill cartridge, another becomes a family heirloom, and another resides in a museum? Ancient philosophers asserted the three motivators of all human inquiry are pursuit of truth, beauty, and goodness. This triumvirate of cognition, aesthetics, and morality offers a framework for looking at design as an attempt to develop functional products and systems, express cultural heritage, and express an aesthetic quality. The first category can be referred to as mechanistic influences. The other categories can be referred to as nonmechanistic (a fancy name for nontechnical) influences such as aesthetics, brand management, and material culture. These nonmechanistic influences can be determined by delving into the social sciences, perhaps most appropriately ethnography, which is the study of human culture.

The Role of Aesthetics in Design

The philosophy of aesthetics is connected with value judgments and is a preeminent concern for industrial designers. Even in a study of engineering principles, this subject cannot be ignored. The mark of a professional designer is integrating aesthetics into everything. Often beauty is subordinated to other concerns such as functionality and brand management, but making a design appealing and intuitive is part of the aesthetic treatment. These are key added values provided by the designer.

Aesthetic can be defined as "pleasing in appearance," and while an important design consideration, this quality is subjective and defies quantification. While three-dimensional forms consider traditional elements of design such as symmetry, repetition, rhythm, balance, proportion, harmony, movement, color, and texture, a broader aesthetic assessment can be informed by all the sensory inputs: sight, sound, touch, smell, and taste. The interplays between these senses are complicated; for example, the visual appearance of a perfume bottle is often as important as the smell, while the thickness, shape, crunch (biting pressure), and color of a potato chip influence the consumer experience beyond smell and taste.

The foundation under which aesthetics are evaluated is a contested issue among philosophers. One philosophical school (aestheticism) believes aesthetic evaluation is independent of other judgments, while others, such as the instrumentalists, believe aesthetic value is connected to function.

A synonym for aesthetics could be appeal. Making a design appealing to the senses has allure. Allure is why all cultures seem to surround themselves with appealing things from nature, whether attractive stones or flamboyant feathers. You might not think of a Hummer as a pretty car but it is appealing in its own way. The same is true with an aircraft carrier or Parisian coffee shop. They have appeal beyond the visual.

The aircraft carrier in Figure 2-1 is probably not considered beautiful—its appeal lies in its capabilities, which are reflected in the ship design. The watch in Figure 2-2 is a rather busy collection of round and linear shapes, yet it has an appeal with its technical proclamation.

FIGURE 2-1. While violating all visual design principles, aircraft carriers have a peculiarly provocative visual appeal

FIGURE 2-2. Complexity can be appealing, as in the case of this watch

A discussion of aesthetics should consider the notion of beauty. Art is often created so its beauty is proffered to human regard. The Austrian-Swiss poet Rainer Maria Rilke's poems speak of beauty making the inanimate alive, where an ancient torso of Apollo "would burst out of its confines and radiate like a star" much like the Jaguar E-Type sports car. Human beauty has its effect: from Helen of Troy's face "launching a thousand ships" seeking to rescue her from Troy to the political intrigue caused by Cleopatra and Antony. The mythological three graces of beauty, charm, and joy (Figure 2-3) entertained the guests of the gods while the poetry of Sappho and Dante drool with notions of beauty and passion. Beauty can evoke protectiveness—you kill a cockroach, but not a butterfly. Beauty is a force to be dealt with, although it is less charged when considering nonsentient objects.

FIGURE 2-3. Detail of the Three Graces from Botticelli's *Primavera*

We drive out of our way to see a beautiful vista or to catch eyes with a charming dog on a porch. We carry things of beauty with us, and like crows, seem to like shiny things that leap out. We often like small things we can wear or hold. Jewelry, watches, wallets, clothes, shoes, and personal electronics are staples of the consumer market. We leave the big stuff to the state to maintain in museums.

Beauty has a historical component and our notion of beauty changes. Time makes human beauty rise and fade from baby to the gruesome corpse. In the same fashion, an object's allure also changes with time. Byzantine paintings may have lost their beauty, but are preserved to maintain a connection with our past. Will Pollock's drip paintings be attractive in 1,000 years? The notion of beauty in the fine arts changes century to century, as illustrated in Figure 2-4.

FIGURE 2-4. These paintings show the changing appearance of contemporary beauty from Byzantine to Renaissance to Abstract

Observing a three-dimensional object has a time component like a cartoon strip. We see different things as we move with respect to the object. This movement shows us new vistas and even surprises like what we might encounter during a walk in the woods. Design that reveals new features, functions, and forms can be equally intriguing.

Some designs are appealing for unique reasons. For example, the sculpture in Figure 2-5 fits nicely into a hand and that has its own appeal. The electronic calendar in Figure 2-6 unfolds automatically—an intriguing and surprising dynamic.

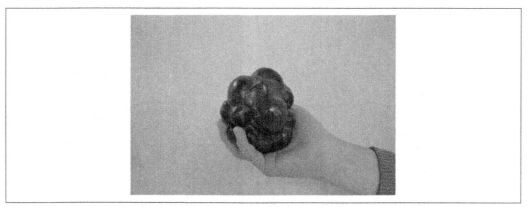

FIGURE 2-5. Part of this sculpture's appeal is that it feels nice in the hand

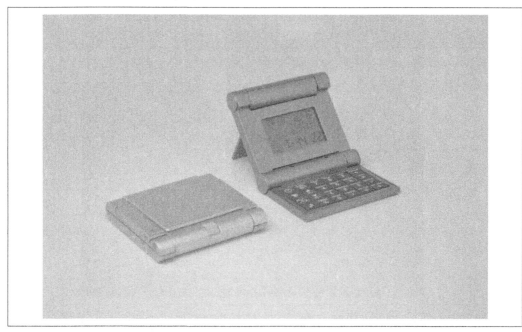

FIGURE 2-6. Watching this calculator unfold is appealing

It is important to note that pretty two-dimensional renderings are not design. Software has allowed us to make a cheese slicer or a lawn mower look like a gleaming pearl with diamond accents. Dramatic portrayals of ideation can show the progression from a cube to an elegant toaster. Software can then make the toaster look like it went to the best beauty parlor in Palm Springs. But is it a good design? Is it appealing in its real "I need to make toast at 6 a.m." context?

There is a difference between the fine arts and design. Fine arts are self-expression, while design is purposeful art that needs to be informed by all the stakeholders and especially the end user. Consequently, when designing for others, aesthetics takes on a tricky character and must be developed not only by the designer's creative senses and style orthodoxy but also by cultural issues that may be best approached ethnographically.

The black utility knife in the background of Figure 2-7 has an inner beauty. It has been carefully designed so that the user can easily open it and change blades. The new blade holder rotates up and makes the whole process pleasant. The red utility knife in foreground requires a screwdriver, patience, and a spare screw after the original one rolls off the workbench and becomes lost.

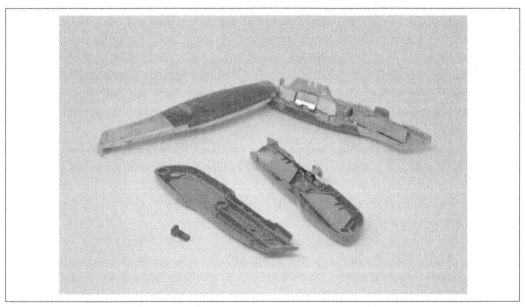

FIGURE 2-7. The (black) knife in the background considers a wide range of design issues: the mechanics of opening, the sound it makes upon closure, the presentation of replacement blades, and the shape of the handle

The Need for Brand Management

Design is not self-expression; rather, it synthesizes a variety of market requirements. One of these requirements is brand management. Designs often need to perpetuate hard-earned brand appeal. The transcendent appeal of branded products is carefully managed and jealously guarded.

As products moved from the local artisan, branding emerged from the marketplace largely to confirm legal protection and guarantee quality as well as homogeneity of product or experience. Branding allows for market control and rapid introduction of new products or features. I remember traveling in India many years ago and buying a boxed set of cookies. Usually goods were not packaged in any fashion and I was eager to enjoy the yummy cookies that the photographs on the box promised. I soon found out that the cookies pictured on the package did not represent what was inside. I was so conditioned to truthfulness in presentation that I didn't think of this possibility. I can't remember the brand but I remember the experience!

Brands add value to a product. This brand equity is an example of customer-perceived value and is worth real money. We don't buy clothes to keep us warm or a car to simply allow us to travel between points. The added values fall into the categories of experience, effectiveness, visual appearance, and social associations.

Brand experience reflects the personality of the brand, such as the comforts associated with Betty Crocker or Birkenstock. The effectiveness of the brand relates to product superiority, whether it is the driving experience of a BMW or the curative powers of proprietary drugs. The visual appearance of labeling can be powerful. Imagine selling unbranded bottled waters and juices at the prices proffered now? Trendy drinks and delicate perfumes live off packaging.

The people associated with a brand are perhaps the leading connection into what is now known as a lifestyle brand. The brand we select represents both our individual and group identity. We value and defend our group/brand. This people association feeds right into natural group dynamics.

Sociological identity theories describing the relationship between an individual and a group are built upon the individual's value for self. Individuals become group members who then need to protect the positive image of their group to satisfy the self-esteem they derived from the group. As a result, they regard other groups as inferior to their group. Group members with strong group identification attributed successful endeavors to the group members and negative group outcomes to external factors. Those who identified weakly with a group were more inclined to attribute negative group outcomes to group members and positive group outcomes to external factors. Myths and stories can also perpetuate a group identity.

To move design beyond individual connoisseurship, you need to consider the following in connection with brand management:

- Themes associated with brand. You could consider action words or moods that are evoked by this brand. Use data (ethnographic and otherwise), not your own personal opinions.
- Verbal associations with brand (e.g., tag line).
- Graphic associations with brand (e.g., logo and colors). How does your design visually connect with the brand?
- Does your design support the brand? Think in terms of:
  - Experience
  - Effectiveness

- Visual appearance
- People associations

Material Culture: The Context of Design

Is the *Mona Lisa* really a beautiful painting? Do we fly to Paris, wait in queues at the Louvre, and then take a picture of the painting through its bulletproof glass structure because we want to cherish the image, or are we more interested in telling our friends about our visit? Perhaps we are trying to connect with others through a communal experience. Perhaps we are trying to connect with an important time in human history or with Leonardo da Vinci himself.

The relationship between the individual, society, and history is considered in the concept of material culture. People understand material objects as they have been learned from their culture and we are compelled to understand the relationship between material objects and meaning. Individuals and groups attribute powerful significance to their history and strive to preserve it. A resistance to design change can derive from the dissonance between the identified cultural heritage and a proposed design change. This resistance can also be understood by considering the power of group identity in preserving the vital definition of a group upon which the individual derives identity. We often critique these designs as being too "something"—trendy, old looking, urban, country, ethnic, commercial, and so on. Design can derive from filial piety, where we identify with our ancestors and honor them by incorporating elements they value into design.

While traditional knowledge can be thought of as taking the past and applying it to the future, material culture motivates people to look at the present and relate it to the past, much like ethnography can inform anthropology and anatomy can guide paleontology.

A sense for material culture can be derived from gaining insights into the culture's history and how the design application you are investigating is typically used. Think about special requirements, such as an oven being oversized to accommodate a Thanksgiving turkey or a refrigerator sized to accommodate a minivan full of food from a supermarket as opposed to a small grocery store. Consider how the product is disposed; does it become an heirloom or garbage? Does the product show up in family photographs and other memorabilia? Do people normally put their name or distinguishing marks on the product? Do people proudly show off the product to others? Is this product suitable for a gift to commemorate a special occasion, such as a wedding? Finally, a designer reflects on how his or her design aligns or offends the material culture.

The Influence of Tradition

Tradition reminds of us of the reliable comforts of home and hearth. We think of traditional knowledge as being refined by time to become an important standard for comparison, from the ingredients in our favorite food, to the tonal structure of music, to the shape of a chair. Traditional or indigenous knowledge is defined by the United Nations Environmental Programme (UNEP) as "the knowledge that an indigenous (local) community accumulates over generations of living in a particular environment. This definition encompasses all forms of knowledge—technologies, know-how skills, practice, and beliefs—that enable the community to achieve stable livelihoods in their environment." Tradition represents one way of doing things or one way of understanding what has been successful in a particular cultural and technological context, providing the underpinning for designs such as used with houses, tools, furniture, and clothing. Traditions are valuable in many respects because they reduce mistakes and reinforce material culture.

Problems with Tradition

Some traditions are deeply embedded in the contours of society, such as road widths or legal systems, while others are new and potentially frustrating, such as increasing reliance upon remote controls, small display screens, or automated telephone systems. Traditional knowledge often supports nonmechanistic requirements such as belief systems, relationship to nature, and aesthetics. However, traditional behavior can be in conflict with normally powerful nonmechanistic forces. Anthropological studies show deforestation, fishing by water poisoning, hunting by large-scale forest burns, burning caches of food for social prestige, or even taunting children into touching fire and laughing at their pain occur even when these actions present some conflict with belief systems.

The problems associated with tradition can be seen in the design community where protective laws or tariffs are enacted or legal entanglements are erected. For example, the antiquated design of Harley-Davidson motorcycles was protected by US tariffs in 1983 and the entrenched automobile industry legally pursued the Tucker Car Corporation in 1949, resulting in part from Tucker's nontraditional safety innovations that threatened the status quo. Within products themselves we see traditional design still giving us slotted screws (versus Phillips), undersized mug handles, excessively wide stovetops, and nonintuitive door operation.

Visual Stereotype: Familiarity Influencing Design

One subset of tradition that can be very helpful in design is the entrenched visual stereotypes we hold. Visual stereotypes are exemplars of form that provide a culturally relevant expectation of function. The visual stereotype helps users understand the product's role. The identified visual stereotype represents an archetype upon which to start a design or as a baseline to compare a new design's nonmechanistic appeal. This information is especially helpful when designing for a foreign culture or otherwise distinct community.

A good example of the power of visual stereotype is in American pickup trucks in which the proportions have remained unchanged for more than 40 years. This continued sense of proportion and visual stereotype is seen when comparing a 1965 and 2016 Ford pickup in Figure 2-8.

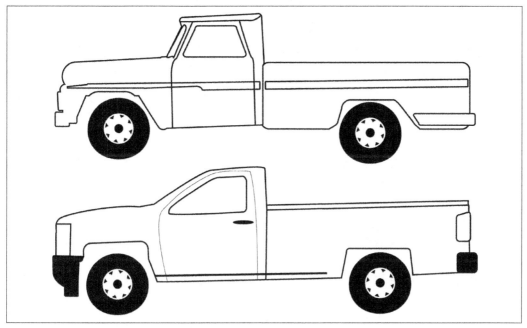

FIGURE 2-8. These visual stereotypes define a pickup truck

Deviations from the visual stereotype can be troubling because the purpose of the design can be unclear and its connection with a particular application (e.g., trucking) becomes confused. A designer must be aware of market expectations and artfully blend the expected form with new and exciting elements. This is much akin to poetry, in which a new language is not created, but rather exists as an established language used in surprising ways.

You can use visual stereotypes in practice by taking photographs of traditional designs and overlaying to find an expected visual stereotype that can provide a helpful starting point for design. For example, obtaining outlines of Malaysian trawlers like those shown in Figure 2-9 can produce the stereotypical profile shown in Figure 2-10. This profile gives us a starting point for design or a baseline upon which to compare design changes. This baseline is a helpful adjunct in ethnographic design.

FIGURE 2-9. Traditional Malaysian fishing boats

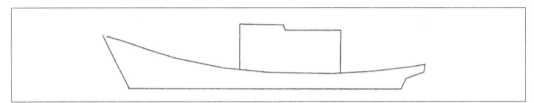

FIGURE 2-10. Visual stereotype of traditional Malaysian fishing boat

Ethnography's Role in User-Centered Design

Ethnography is the study of human culture and strives to extract truths based on holistic and richly detailed subjective appraisals of small populations. Ethnography is an active form of research that requires the researcher to obtain data by observation or interviews while being able to respond to variables such as changing mood of the respondents, nonverbal behavioral cues, and sensitivity to ethical constraints. All the while, the ethnographer must be aware of this influence on respondents' behavior and how it might skew results. Moreover, ethnography requires data triangulation and rigorous treatment of data in order to obtain valid conclusions.

Ethnographic design most commonly employs interviews and observation of potential users. These same techniques can also be applied to other stakeholders, such as retailers, repair services, and so on. Sometimes, potential users can give you information in creative ways. Figure 2-11 shows a Malaysian boat builder using clay to describe boat form and technical features.

FIGURE 2-11. Boat builder carving clay to communicate design details

Some of richest ethnographic data is derived from participant observation. This is where a trained ethnographer joins a group, and intimately knowing their social perspectives, observes how they interact with a design. However, focus groups and a variety of interviewing techniques are often used in industry, and they can provide a deep and focused insight into a design and its socially derived meaning.

Ethnographically driven design has the advantage of broad application and allows a designer to work outside his or her culture to create effective designs. While it is unfair to ethnographers to say you can just read some books and go out and "do ethnography," if ethnography is closed to designers, exclusionary barriers are raised that cripple both design and ethnography. The crippling is caused by failure to consider data that can only be approached ethnographically and depriving ethnography of practical application. We need to feed from a deep trough for our designs—physical sciences, life sciences, math, anthropology, sociology, psychology, and economics can all be tapped for methods that improve design. Designers need to borrow from all disciplines with alacrity and continually explore. Figure 2-12 offers a structure of culturally relevant design.

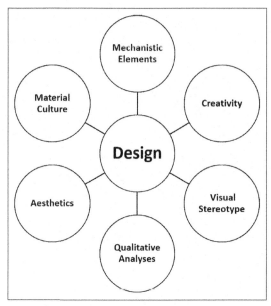

FIGURE 2-12. Designs derive from many inputs

Helpful Art and Design Vocabulary

General Design

Mechanistic
  Mechanically determined, explaining phenomena only by reference to physical causes.

Nonmechanistic
  Everything else: aesthetic, material culture, visual stereotype, and so on.

Bounded rationality
  Recognizes that decisions need to be made based upon partial or fragmentary information.

Visual stereotype
  Expected form and proportions of an object, such as a pickup truck.

Context
  The physical, virtual, and social structures that surround the point of use. For example: home field advantage in sports, dinner table conversation versus cafeteria conversation, showing off new electronic gadget in public.

Affordances
  Suggest how you can deal with a device (e.g., holes in a scissor handle).

Constraints
  Limits on how a device can be used (e.g., size of holes).

Mapping
  Relationship between action and results (e.g., moving fingers, moves scissor cutting edges).

Cultural constraint
  Culturally derived (e.g., red stop lights).

Physical constraint
  Mechanical limits (e.g., a key operating a lock).

Semantic constraint
  Logical sequencing (e.g., locating a windshield in front of a car driver).

Learned helplessness
  User confusion and mistakes arising from faulty mental models.

Informant
  Person who provides verbal information, such as through an interview.

Respondent
  Person who responds to some query, such as through a survey or questionnaire.

Exemplar
  Informant response that is quoted and provides an example of findings.

Material culture
  People attributing a cultural identity to material things.

Cultural heritage
  Legacy of physical artifacts and the recognition that individuals and groups attribute powerful significance to their history and strive to preserve it.

Art and composition

Shape
  Two-dimensional line with no form or thickness. Shapes are flat and can be grouped into two categories: geometric and organic.

Form
  Three-dimensional object having volume and thickness. Form can be viewed from many angles.

Texture
  Surface roughness either tactile or visual and can be real or implied.

Proportion
  The size, location, or amount of one element to another (or to the whole) in a work.

Balance
  Visual equality in shape, form, value, color, and so on.

Harmony
  Consistency in a design, such as the consistent use of organic or geometric features.

Directional movement
  Visual flow produced by a design's lines or values that can suggest motion in a design.

Rhythm
  Regular repetition of some design feature.

Light and color

Value
  Degree of light and darkness. Black has a high tonal value, while white has a low tonal value; grays fall in between. Value can be used with color as well as black and white. Contrast is the extreme changes between values.

Color
  Color has three properties: (1) the traditional color name, called hue (e.g., red), (2) the lightness of the hue, and (3) the brightness of the hue.
  The lightness of the color called value or brightness is used to differentiate a light blue from a dark blue. The brightness of a hue is sometimes referred to as saturation and indicates the color intensity. A hue is most intense when it is the pure color as might appear on a color wheel. As it is mixed with other colors it becomes less intense. To add to the confusion, all of these terms have synonyms, such as saturation, chroma, and luminance. In practice, think of a color being produced by taking a hue from the color wheel, adding black to develop a value, and then using a complementary hue to make it less intense.

Opaque
  Having covering power; not permitting light/color to penetrate.

Translucent
  Permitting partial light/color penetration.

Transparent
  Permitting complete light/color penetration.

Primary colors
  Red, yellow, and blue. With these three colors (and black and white) all other colors can be made. The primary colors themselves cannot be made by mixing other colors.

Secondary colors
  Orange, violet, and green. These colors are created by the mixture of two primary colors.

Analogous colors
  Adjacent colors on the color wheel.

Complementary colors
   Opposite colors on the color wheel. The complement of red is green, the complement of yellow is orange, and the complement of blue is violet.

Warm colors
   Red, orange, yellow, which tend to advance in visual space.

Cool colors
   Violet, blue, green, which tend to recede in space.

Tint
   Mixing a color with white.

Shade
   Mixing a color with black.

Creativity: The Fount of Design

Because design is a practical application of creativity, it is worth a moment to consider some foundational aspects of this ability. Creativity is an abstract quality that can quietly aid our pedestrian problem-solving abilities or engorge our senses in artistic expressions. Creativity is most often recognized and honored in the arts and sciences, and a "creative person" is commonly associated with someone capable of producing art, literature, or scientific innovation. Creative expressions in other forms are not provided similar accolades. Therefore, we can't discern between creative versus noncreative people in a universal sense, if that delineation even exists, by societal judgments. Shakespeare's plays are still being performed, thereby reinforcing their relevance to contemporary Western society, whereas the game devised by some parent in a remote village disappears into the flow of time.

Investigating individual creativity may also have a disruptive effect on the process itself. The Swiss psychiatrist Carl Gustav Jung noted "as long as we ourselves are caught up in the process of creation, we neither see nor understand; indeed we ought not to understand, for nothing is more injurious to immediate experience than cognition." However, we can gain insight into individual and group motivations for acting creatively. These encouraging forces can best be considered by investigating different psychological and sociological theories for behavior.

While psychological theories describe individual motivations for creativity, our desire to act creatively is also influenced by social context. Groups can support or hinder creativity. Creativity can be part of a culture or dismissed as a distraction. The influence of groups upon creativity can be quite overt, as in the employment of scheduled brainstorming meetings, or the use of tangible rewards for creativity, such as money. However, the group forces can also be subtle. A group can view itself as being creative and therefore the individuals within the group also consider themselves creative. This self-identification is similar to the brand appeal arduously earned by sophisticated advertisers and is highly potent in directing individual behavior.

The best description of the nature of creativity lies in a mix of psychological and sociological theories that attempt to describe both the individual motivations for creativity and the influence of group behavior. The Norwegian artist Edvard Munch, who suffered from depression, paranoia, and many phobias, said: "My affliction belongs to me and my art—they have become one with me. Without illness and anxiety, I would have been a rudderless ship." This unfortunate relationship shows us that creativity can be associated with powerful cognitive forces—a dance between a primal, perhaps slightly mad, cognition, and a consciousness that is needed to tame it.

Creativity often requires a combining of two ideas with related functions, such as adding a claw for removing nails to the back of a hammer. It also may require the development of the desired goals or market demand to provide a focus upon the thinking process. This latter notion is reinforced by the American physicist Arthur Schawlow, who claims "the most successful scientists often are not the most talented, but the ones who are just impelled by curiosity.... They've got to know what the answer is." Creativity, when viewed from many perspectives, can be seen as combining two or more ideas, processes, or daydreams. Therefore, the collaborations among people multiply the opportunity to combine ideas.

Creativity often seems to prosper in a collaborative environment. The connection between collaboration and creativity appears to be an innate need. The American psychologist Howard Gruber noted "Creative people must use their skills to devise environments that foster their work. They invent new peer groups appropriate to their projects. Being creative means striking out in new directions and not accepting ready-made relationships.... Each creator therefore invents new forms of collaboration." According to the American anthropologist Edward T. Hall, "There is a natural drive to collaborate. It may be one of the basic principles of living in substances."

The interactions of people within a group are an important component of creative expression. Creativity can be nurtured or nullified in a group. While Picasso and Braque collaborated to lead the Cubist movement and Einstein worked with Grossman to develop the mathematical language of nonlinear geometry for expressing relativity theories, many creative people from Sappho to Shakespeare did not collaborate. But how many great creators have had their creative products discarded when their life ended? How many more individuals who, while working within a group, had a wonderful idea attacked or ridiculed so as not to ever be developed? We will never know.

The nature of creativity lies among various psychological and sociological models. While these theories do not predict actions or offer any complete description of creativity, they offer a fragmented foundation for gaining insight. Is creativity an innate talent or something drawn from us by other forces or circumstances? The individual must be creative in order to manage daily challenges, but creativity can arise in spectacular form in the arts and sciences. Creative collaborations can lead to dramatic objects of art and literature as well as powerful changes in human perspective. However, these are forms of creativity that are rewarded and recognized and do not represent the range of human creative expressions. Games with a child or daydreams may express as much creativity as Dante's *Inferno*, but it is not recognized in the same light. Moreover, some people never have an opportunity for their creative expressions to find a public forum. Therefore, the individual propensity for creativity in a general sense is impossible to assess.

Sociological models of individual self and its relations to groups indicate people benefit from group participation and identification. The group you identify with can have consequences beyond the functioning of that group—the group can in turn define your self-worth. Moreover, this relationship between self and group can adversely affect your view of those outside your group. Identity theory suggests that when self-proclaimed creative people gather in groups, they will deeply nurture one another's creativity and at the same time excoriate other groups' creative efforts. We can see in reviewing historic collaborations of artists and scientists that they gained confidence based on numbers. Therefore, while individual creativity is difficult to appraise, a group culture can have a predictable effect upon the individual members' creative expression.

What Is Creativity?

Creative action normally has to pass three tests: it must be original, valuable, and require special abilities. Like poetry, creativity often embodies something new and unexpected. Judging a product's creativity can be based on newness, effectiveness, and the manner in which the problem is solved. These basic categories are referred to as novelty, resolution, and elaboration and synthesis, respectively.

A product with low novelty in reproducing a form might be a chisel, while a product with high novelty might be a 3D printer. This rating would be akin to the low novelty of Norman Rockwell's art versus the high novelty of Salvador Dali's art. In the resolution category, a fastener product with low resolution would be a button in contrast to the mechanically complex zipper. This is like comparing blank verse poetry to a structured sonnet. Finally, if elaboration and synthesis were considered for a mathematical machine, an abacus or slide rule would be rated low, while a computer would be rated high.

This rating system has problems in actually attributing creative value, and a high ranking in each of these categories does not necessarily dictate creative superiority. However, these categories provide a helpful language in critiquing designs.

A reflection on what creativity is could be simply anything that provides a delightful surprise.

Creativity Exercises

The following can compel you to think and act in uncommon ways. These exercises represent activities that can be founts of creative expression:

1. Draw an image with the following constraints: no representational art, use abstractions and color. You can use some imagery as you see fit but make it more an expression of emotion and visual impact. The depiction should be of an abstract noun such as honesty or ambition. Use lines, curves, value, and texture. Use colored pencils, pastels, or whatever you like.

2. Explain a painting in words alone. Good paintings for this exercise are Monet's *Impressions at Sunrise* or Munch's *The Scream*.

3. Write a short poem about an abstract thought, such as how a beautiful sunset makes you feel. Free verse is OK.

CHAPTER 3

Material Mechanics

DESIGNERS WORRY THEIR DESIGNS WILL BREAK. WE MUST HAVE A DEEP, INTUITIVE SENSE FOR the behavior of materials so that among the flourishing ideation, we don't end up with designs that can never work in the real world. Therefore, we need to understand the behavior of solid objects when a load is applied so we know whether it will break or bend to the point where it can't do its job. Designers also need to understand how shape affects the rigidity of an object, how much it will bend or twist when it is being used. While the theory of material mechanics must not intrude on the creative process, it must hold a place in refining ideas so products are not designed to break. While Louis Sullivan's aphorism that "form should follow function" is not always true, there exists design chauvinism when designs flaunt the laws of nature. Flaunting of nature can be appealing, sometimes suggesting motion and instability. It can have artistic merit. The goal of this section is to understand material behavior so we have a sense for how our designs will work in real-world applications. Like the fine arts, the designer should sit in the driver's seat and not just hope an engineer will rescue your design from failure.

We want to understand how a solid object responds to being used. We will look at the relationship between stress and an object's shape as well as sneaky things such as fatigue fractures, buckling, and thermal expansion. These insights will allow you to synthesize these concerns while in the early stages of design and creative exploration.

 Tools

This chapter considers solid materials, but we will look later at liquids and gases. While this theory should inform your design, this reading must be coupled with experimentation. You want more than a technical vocabulary and scientific certitude. You need to break stuff. You need to bend and twist things—all sorts of things. Grab some rulers, pencils, soda straws, and paper clips and mess around.

The Effects of Pulling, Pushing, and Twisting Forces

Solid objects move when a force is applied to them. If they are constrained, they may not move in some way you can see. However, they will get squeezed or stretched internally. The force generates internal stresses that can cause the material to break. The way an object moves internally is the key to understanding a material's behavior under load.

Every type of material has a unique way of stretching or compressing. Some materials are viscous, like a fluid; others are elastic, like a solid. Some materials fall in between, such as plastic and blood. They flow like a liquid yet can be stretched and recover to their original dimension like a solid. These materials are called *viscoelastic* and will be discussed in Chapter 6. The relationship between the force produced by a load and the movement of the material defines how it can be used.

Although a force can only produce tension, compression, and shear, it is convenient to think of loads being applied to an object in five ways:

- Tension
- Compression
- Shear
- Torsion
- Flexure

Tension is simply a force that pulls apart an object's molecules, while *compression* is the opposite of tension and is the squeezing of an object's molecules. Most materials have a higher compressive strength than tensile strength.

Shear is a set of opposing forces in the same plane. For example, if the bottom of a plate is rigidly attached to a surface and then the top of the plate is pushed parallel to the surface, a shear force will be applied to the plate. The shear force occurs when the force at the bottom of the plate resists the force on top of the plate. These opposing forces are transmitted as a shear force in the plate. If the plate were made from a stack of papers, the shear force would readily allow the papers to slide across one another. The ability to bear a shearing stress is the feature that defines solids. Solids resist shearing forces while fluids (gases and liquids) will flow until the shearing stresses are gone.

Torsion is a twisting force. For example, a rod that is held rigid at one end while a rotating force is applied at the other end will experience a torsional force on the rod. The outer surface has the highest stress and the centerline will have zero stress. The behavior of a design can be very different when subject to torsional loads as opposed to flexural or other loads. It is for this reason cross bracing is used in many structures. For example, a box-shaped structure like a bookshelf can have its torsional deflection reduced 100 fold by including cross bracing. Simpler diagonal bracing is also very effective, but this design generally permits three times more deflection than cross bracing. Torsional rigidity can also be improved by incorporating closed members such as tubing rather than open members such as U channels.

Torsion is important to understand in a variety of applications, such as the bookshelf previously mentioned, but also propulsion systems where all the power is transmitted by torque. Not only can this torque tear apart shafts but it can produce torsional vibrations that need to be stopped by changing material properties or with dampeners. Helmets must be designed for more than impact loading; they must reduce the twisting of the head. This can be done with membranes that allow a sliding action to occur between the shell and the head.

Flexure is a load that does not fall in line with the central axis of a part, like standing at the end of a diving board. This is an extremely common loading condition and is produced by an offset load that creates tensile loading in one half of a part and compressive loading in another half. If a bar is supported at both ends and a load is applied in the center, the material on the same side of the centerline as the load will be in compression whereas the material on the opposite side of the load will be in tension. In addition, the stress varies with the distance from the centerline. Flexural stress is harder to understand than the other stresses because it is a function of a part's shape. For example, an I-beam is much stronger in one plane than an equivalent mass of square steel. This will be considered in more detail later.

Tension, compression, shear, torsion, and flexure—now get a paper clip, stick, or a ruler and try them all out. Pull, push, shear, twist, and bend and see how it behaves. Notice the ruler twists pretty easy. It is rigid in one axis but it can easily be twisted and bent in other axes. Now try twisting a pen. It is torsionally rigid. What appear to be strong objects can be very weak in some stress conditions. You can park a car on a pile of papers, but they are easy to shear or pull apart. While this will all be good fun, you will need some squishy clay to really see the effect of compression. Figure 3-1 shows a summary of these loading conditions.

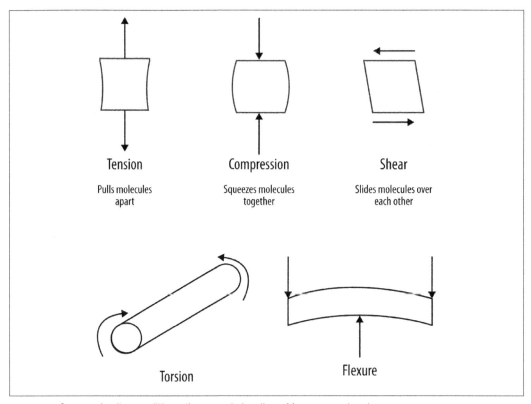

FIGURE 3-1. Common loading conditions; these create tensile and/or compressive stresses

Why does a designer care about this? Because only you know what loads might be applied to your design and you must understand the stresses these loads will produce. We will discuss later how the shape of your design will dictate the intensity of these stresses and therefore how strong and rigid your design is. These terms have different meanings: strong means your design won't easily break; rigid means it won't easily deform.

Static Loads

When considering the behavior of a design, we typically start with evaluating how loads and twisting action are transmitted through an object. Loads move objects. Either they accelerate and move or they make small internal movements called strain. If you are sitting on a teeter-totter with someone of equal weight on the other side, the teeter-totter is balanced in a state of equilibrium. If you push up with your legs, that force propels you up and the other person down until they push back on the ground. If you extend a ruler over the edge of a desk and push down on it, you can feel the load in the hand holding the other end. If you push down further from the edge of the ruler, the resisting force you have to apply increases. The reaction to a force and a distance is called a *moment*. Moment is also referred to as a *couple* or *torque*.

Equilibrium requires that the summation of all forces and moments on an object be equal to zero. This study of how forces transmit through a stationary object is called *statics*.

The term *load* can be used loosely—when we stand on a floor, we are putting a load on it. However, the term describes two things: force and direction. In the case of the floor, the force is your body weight and the direction is down in line with gravity. The loads and moments upon which an object is subject depend on both the magnitude of the load and its distance from a support. For example, the deflection of an overhung (cantilever) beam is sixteen times that of a beam simply supported at both ends. You can experiment with a ruler and different supports to get a real sense for this behavior. This perception of loads is intuitive for us because our limbs are a bunch of beams that are usually cantilevered. We know it is hard to hold something far from our body. We prefer our limbs to be supported at both ends, hence we like to put our elbows on the table while we are eating (a fork full of food is heavy!) so at least our upper arms are supported at both ends.

When a force is applied to an object, it follows Newton's three laws of motion. The first two relate to dynamics. The first law is that every object in a state of uniform motion tends to remain in that state of motion unless an external force is applied. This corresponds to the inertia of an object. Inertia is a property of matter and matter does not want to accelerate or decelerate on its own—it needs a force to make that happen.

Newton's second law states that force is equal to mass times acceleration. Therefore, it takes force to accelerate a mass. The higher the mass or the higher the desired acceleration, the greater the required force. If you put a model rocket engine on a car, nothing interesting will happen, but if you put the same engine on a much less massive model rocket—wow!

Newton's third law brings us right back to where we started with regards to statics. The third law states that for every action there is an equal and opposite reaction. When you stand on a bridge, the bridge is pushing up on you with the same force. You will have noticed this if you try stepping off a canoe. When you push off the canoe, the canoe can't push back very well because it has low inertia and is floating on water so it gets pushed away from you.

Identifying all the forces with which an object must contend is necessary in evaluating a design. Some of these might not be obvious, such as the mechanic who uses a filter bracket as a footstep to service a turbocharger. The second biggest load this bracket would encounter is when the engine is being transported by rail and the cars smash into each other during latching. Think about the damage caused by the expansive force of ice in roads, or bird droppings on a car. Toyota recently recalled more than 800,000 cars because of the potential for a spider web to block the condensate drainage of the air conditioner and damage the air bag electronics.

Many forces in nature are trying to destroy your design in a beautifully natural way. The design world is full of these unanticipated usages and they must all be considered. Toy designers probably have the biggest challenge in this area—kids do amazing things!

Draw upon your creative insights, and think about how the object could be misused and what this might do to the object. Are screwdrivers only used to drive in screws? Of course not! Figures 3-2 and 3-3 show examples of misuse.

FIGURE 3-2. If you design doorknobs, doors, or door hinges you need to know about this traditional application

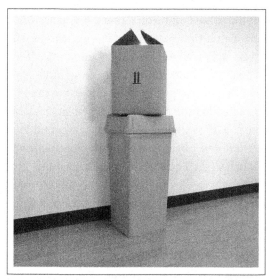

FIGURE 3-3. Garbage cans are not just for garbage; they are great places to rest a box. How heavy a box should it hold?

Forces are normally sketched out on free body diagrams, such as shown in Figure 3-4 and Figure 3-5.

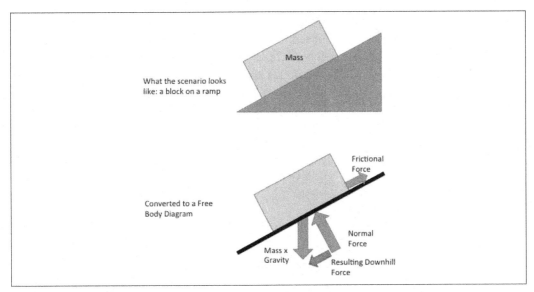

FIGURE 3-4. Free body diagram

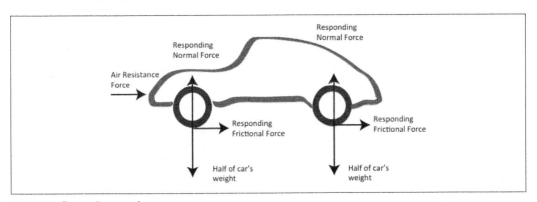

FIGURE 3-5. Forces diagram of a car

This diagram indicates the load, direction, and type of force. Remember to consider the loads, temperatures, and other environments encountered during shipping, handling, and expected misusage as previously described.

Mechanics of Deformable Bodies

In addition to the reaction of an object to a load, another important characteristic of a material is its behavior as it is about to break. Designers care about how their products will stretch and deform because we understand the weak points and vulnerabilities of a product more than anyone else. Some materials, like glass, are brittle and hardly stretch before they break. Other materials, like wood, are ductile and stretch noticeably before they break. This ductility characteristic determines the existence of common warning signs (bending or stretching) prior to imminent failure. The category of deformable bodies includes all solids because they can develop internal stresses. The behavior of fluids will be described in Chapter 6.

A force's effect on a material is dependent on the size of the area upon which the force is applied. The force per area is called *stress* and is simply equal to force divided by area:

$$\sigma = F/A$$

where:

σ = stress
F = force
A = area

Strain is the same as stretch but refers specifically to the movement per unit length. Stress-strain diagrams have been developed for most materials by taking a sample of the material and applying force while monitoring its length. The force is continuously increased until the sample breaks.

Most materials you encounter are elastic. That is, over a certain range of stresses the strain is directly proportional to the stress (*Hooke's law*). Moreover, when the stress is removed, the materials return to their original unstressed dimensions. Flexing of a material can be part of a compliant mechanism such as with springs and paper clips.

Both rubber bands and steel rods act like an elastic material. If supporting the same weight over the same cross-sectional area, a rubber band will stretch much more than steel. This difference in stretch is described by the slope of the stress-strain line and is called the *modulus of elasticity* (or *Young's modulus*) and is equal to the difference in stress divided by the difference in strain:

$$E = \sigma/\varepsilon$$

where:

E = modulus of elasticity
σ = stress
ε = strain

The unit for force is pound (lb) or Newton (N) in SI units. The units for stress are pounds per square inch (psi) or Pascal (Pa) in SI units. One Pascal equals one N/m^2. The conversion is: 1 psi = 6,895 Pa. Strain is measured in length per original length—this gives a strange unit of in./in. or mm/mm. Because the modulus of elasticity is stress divided by this unit canceling strain, the modulus of elasticity becomes psi or Pa. Table 3-1 shows the relationship between stress and strain for different materials before they break or permanently stretch.

TABLE 3-1. Moduli of elasticity at 71°F (21°C)

Material	x 1,000,000 psi	GPa
Osmium	80	552
Steel	30	206
Brass	16	110
Aluminum	10	69
Concrete	3	9.2
Wood	1.5	4.6
Polyethylene	0.014 to 0.18	0.096 to 1.24
Rubber	0.0006 to 0.50	0.004 to 3.4

What happens to a material as the stress is continually increased? The *yield strength* (or *elastic limit*) is reached, beyond which the material continues to stretch but will not completely recover if the stress is removed. This permanent stretching (called *yielding* or *plastic strain*) at stresses above the elastic limit occurs in what is called the *plastic region*. As the stress is further increased, the material finally breaks at a load called the *ultimate strength*. Figure 3-6 shows the relationship between stress and strain for ductile and brittle materials.

Where does the material for the stretching come from? In the elastic region, the stretching comes from the increased distance between atoms. In the plastic region, groups of atoms (crystals or grains) slip or deform to allow the stretch. As the material strains in one axis, it also strains in the perpendicular axes. This intuitive relationship is denoted by the ratio of strains called the *Poisson's ratio*.

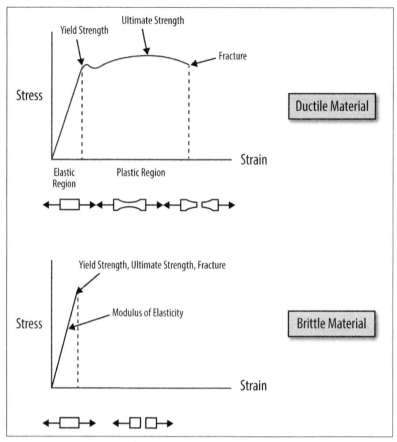

FIGURE 3-6. Stress-strain diagrams for ductile and brittle materials

Consider these practical implications of the relationship between stress and strain:

- Usually the most important stress level to consider is the yield strength rather than the ultimate strength. That is, all considerations of safety factors, fatigue strength, and impact loading are based on a value that will allow the material to completely recover from the applied stress. Notable exceptions are body armor and crash cushions where the energy of the impact is absorbed by taking the material up to or beyond its ultimate strength.
- The modulus of elasticity determines the flexibility of a material and is independent of its strength. However, an object's size and shape is the most important determinant of its flexibility. For example, a thin stick blows readily in the wind whereas a thick trunk made of the same wood does not move at all. The relationship between shape and rigidity is presented in more detail later.

- The area under the stress-strain curve indicates a material's toughness or energy absorption ability. The larger the area, the more energy the material can absorb before it breaks. For example, plastic may be weaker than ceramics but it is tougher.
- Unlike metals, typical composites like wood, fiberglass, and carbon fiber do not have the same stress-strain characteristics in all axes (known as *anisotropic*); therefore, the modulus of elasticity and yield strength will be different depending on orientation. If a load is pulling in the same direction as the fiber orientation, it will be much stiffer and stronger than if it is pulling perpendicular to their orientation. With a perpendicular load, the fiber strength is unused and only the bonding material (such as epoxy) offers resistance to the load. Metals are *isotropic* materials and have the same stress-strain relationships in all axes.

Shape

Before considering how an object's shape affects its strength and rigidity, let's recall that forces produce some combination of tension or compression. However, it is helpful to consider forces present in five different categories. As described previously, these are tension, compression, shear, flexural, or torsional loads.

Shear loading is sliding force whereby the molecules are being asked to slide against each other as opposed to being pulled apart as with tension or squeezed as with compression. In the case of fluids, the shear resistance is very low so it flows easily. The stresses produced by tension, compression, and shear are the easiest loads to calculate because they are simply calculated as force divided by cross-sectional area. Therefore, increasing the area of something, say using a 1/2-inch (12 mm) bolt instead of a 1/4-inch (6 mm) bolt, decreases the stress. Thick is good (but heavy).

RELATIONSHIP BETWEEN SHAPE, STIFFNESS, AND FLEXURAL STRESS

Strength and stiffness come from two different features. Strength is a property of a material while stiffness is a property of the shape of the material. A piece of sheet metal is strong enough to withstand a high level of stress before breaking, yet it easily flexes and buckles. To stiffen the assembly, the sheet metal could be layered upon each other to produce a thicker assembly. This thick assembly of sheets of metal would be rigid and strong but heavy and expensive.

In metal and wood construction, the sheets are stiffened with a framework. In fiber-reinforced plastic (FRP) sandwich construction, an assembly combines a thick layer of low-density material (e.g., foam or honeycomb material) bonded between an inner and outer FRP laminate. The resulting sandwich composite is lighter, less costly, and nearly as strong and as rigid as an equally thick laminate.

RELATIONSHIP BETWEEN STRESS AND THICKNESS

Corrugated cardboard is amazing material. Stiff, strong, and lightweight. This cardboard construction demonstrates an example of sandwich construction. This corrugated cardboard is made entirely of paper with a wavy (corrugated) section of paper sandwiched between the outer layers of flat cardboard. The corrugated cardboard is made rigid because the corrugations separate the flat cardboard, which does not stretch easily, away from the neutral axis. Imagine how floppy the cardboard would be if the corrugated section was flattened out and all the paper layers were laid on top of each other. Even if the flattened paper were replaced by thin sheet metal, it would still not be as rigid as the corrugated form. This increased rigidity and strength are only apparent when the composite is bent. In pure tension, it has virtually no effect—that is, the flattened paper would be as rigid as the corrugated paper if it were pulled at the ends. Figure 3-7 shows a relationship between thickness and stress.

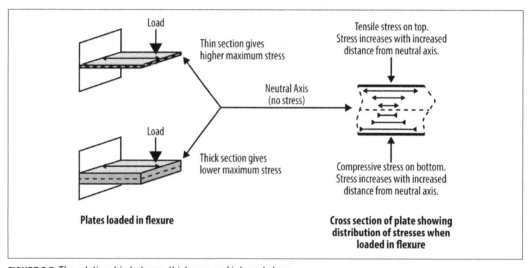

FIGURE 3-7. The relationship between thickness and internal stress

The relationship between rigidity and shape is described and quantified by the *moment of inertia*. Moment of inertia is specifically related to the shape of the cross section under load. In the case of flexural loading (a load that produces only a moment, not an axial or torsional force), the stress produced in the part is inversely proportional to the moment of inertia:

$$\sigma = \frac{Mc}{I}$$

where:

σ = flexural stress
M = moment on section equal to the force times distance
c = distance from neutral axis (usually, the only value of concern is the maximum c value, which is half the thickness of the section)
I = moment of inertia

The higher the moment of inertia, the lower the stress produced under a given load. If a part is designed so the material is far away from its center, or neutral axis, it will have a higher moment of inertia than if all the material is bunched up around the center. In fact, the moment of inertia increases as a cubic function of the thickness. Figure 3-8 shows how a beam can be made 75% more rigid by changing its shape.

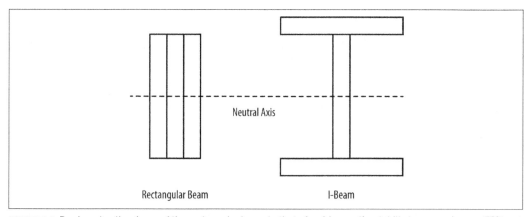

FIGURE 3-8. By changing the shape of the rectangular beam to that of an I-beam, the rigidity increases by over 75%, yet the weight is unchanged, however, this is only true in the vertical plane; in the horizontal plane, the I-Beam is 250% less rigid than in the vertical plane; this illustrates that if the direction of loading can be predicted, it is very beneficial to develop a shape that puts as much of the material as far as possible from the neutral axis

For a rectangular cross section, the formula for I is:

$$I = \frac{bh^3}{12}$$

where:

b = width of the section
h = thickness (or depth) of section

For a solid rod cross section, the formula for I is:

$$I = \pi r^4/4$$

where:

r = radius of rod

A tube has a much higher moment of inertia than a solid rod of the same mass because the mass of the tube is located further from the center. Figure 3-9 shows two round shapes that have the same rigidity even though the tube is almost one half the weight of the rod. Unlike a tube, an I-beam is optimized for resisting bending in one plane. Therefore, a load must be positioned in a way that takes advantage of its shape.

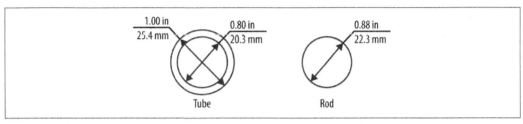

FIGURE 3-9. Tube versus solid rod

Section modulus is often used to describe the properties of a structural shape when dealing with structural steel. Section modulus is equal to the moment of inertia divided by the distance from the center of the shape to its furthest edge (c), such as the radius (r) or half the height (h) in the case of a rectangular shape:

Section modulus (S) = I/c

Torsion has a similar relationship except the design's polar *moment of inertia*—that is, the shape from a twisting perspective—dictates the torsional shear stress. For example, a large-diameter pipe with a thin wall will have a higher polar moment of inertia than will a small-diameter pipe with a thick wall. Torsional shear stress (τ) is equal to the torque (T) times the distance from neutral axis to outer fiber (r) divided by the polar moment of inertia (J):

$$\tau = T r/J$$

where:

  τ = torsional shear
  T = torque
  r = radius from neutral axis to outer fiber
  J = polar moment of $\pi R^4/2$

Few items are under pure tension, compression, or shear. Some items, such as bolts, are designed to be under pure tension and shear only, but these are the exception. Consequently, understanding flexural behavior is essential for predicting potential failures.

Flexural behavior does not consider another phenomenon that will make parts fail at stresses well below their yield stress. This phenomenon is called *buckling* and is a common failure mechanism for long, slender objects. Buckling behavior, a well-understood phenomenon, is carefully considered in design calculations.

Stress Concentration

If you want to know where your design is most likely to break, look for sharp corners and notches. A sudden change in an object's form, such as a sharp corner, notch, hole, or crack, will result in an increase in stress called a *stress concentration*. These geometric changes tend to squeeze (concentrate!) stresses into a smaller area, as shown in Figure 3-10. Stress concentrations also arise around microscopic fissures, which can lead to fatigue failure that will be described later. The mathematical analyses of stress concentrations are so complex that their amplifying values are obtained experimentally.

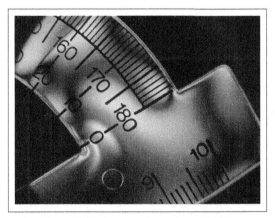

FIGURE 3-10. Stress concentration shown in sharp corner of flexed protractor by photoelastic visualization

Figure 3-11 shows a graph of stress concentration versus radius sharpness. This relationship between shape and stress concentration illustrates the importance of gradually bending mating surfaces. Cracks concentrate stresses.

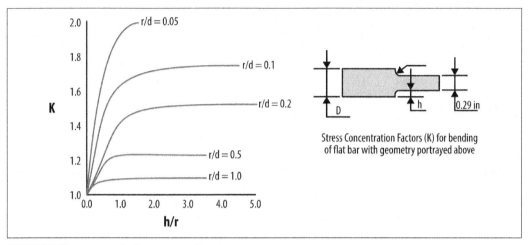

FIGURE 3-11. Stress concentration versus radius sharpness

Stress concentration effects are reduced by moving the abrupt change of form to a lesser stressed section of the object. The problem of stress concentrations has long been known, and boat builders would seek trees with natural fillets for use in highly loaded sections such as keels and mast connections. These valuable forms are stockpiled, as shown in Figure 3-12.

FIGURE 3-12. Wooden sections saved for highly loaded features in boat construction

In building construction, holes are drilled in the center of beams where very low stress exists (see the section "Relationship Between Shape, Stiffness, and Flexural Stress" on page 43) and away from the area of maximum moment. It is accepted practice in some applications to drill small holes at each end of a crack to reduce the stress concentration in this area and to stop the crack's progress.

Ductile materials are less sensitive to stress concentrations than brittle materials because they deform when loaded. This reshaping redistributes the load over a broader area. Even though the brittle form of a material can often handle higher stresses, a ductile material can distribute stress and therefore actually handle a higher load.

Fatigue

As a material is continually loaded and unloaded, its maximum strength decreases. This gradual weakening is called *fatigue*, a name derived from an early misunderstanding of this phenomenon. The metal was thought to have become "tired" over time. Fatigue is now understood to be caused by movements within the material that cause cracks to propagate through the material's grain boundaries. One solution to fatigue failure is to over-design an object—that is, use massive thicknesses to ensure longevity. However, understanding the rate of decline of a material's strength is a more environmentally and cost-sensitive solution. Generally, if the stresses experienced by an object under cyclic tension and compression loading are less than one-half of the yield stress, the object will never fail due to fatigue. Therefore, highly efficient designs are produced only when the anticipated loading and fatigue characteristics are fully understood.

Fatigue failures often start at a stress concentration or at the surface of an object, where stresses are highest. Small surface cracks may be the first sign of fatigue. However, there may be no visible cracking because the cracks may be too small to see, start from inside the object, or be obscured. For example, the stress concentration at the corner of a notch in a shaft (such as the keyway) is a common location for fatigue cracking but this area cannot be observed unless the shafted assembly is completely torn apart.

Inspecting a broken object that is subject to fatigue failure will always show two different surface characteristics:

- The *fatigue zone* where cracks slowly propagated into the object
- The *instantaneous zone* where the crack accelerated quickly through the object

The relationship between the size of the fatigue zone and the instantaneous zone indicates the relative loading of an object. The instantaneous zone is produced when the effective area (total area minus fatigue zone) is too small to support a load. Therefore, if the instantaneous zone is small, the object was not heavily loaded. However, if the instantaneous zone is large, the object was heavily loaded.

The fatigue zone is identified by *beachmarks*, which are small ridges on the broken surface of a material. These beachmarks are produced by the crack propagation. They in turn plot the crack progression. Due to abrasion within the crack, older fatigue zone areas are smoother than new ones and can help identify the failure origin.

FATIGUE STRENGTH

Fatigue strength describes the highest stress a material can handle for a specified number of cycles. Typically, if you want to have a part last an infinite number of stress cycles, the fatigue strength is half the yield strength. That is, an object that will be vibrating or encounter many loading cycles should only be designed to handle half the stress of an object that does not encounter many stress cycles. Simply put, if something experiences a lot of stress cycles, make it twice as thick (which is one way to double the moment of inertia). Again, this is a simplified statement. The actual calculation for required moment of inertia is more complicated.

Although *fatigue strength* is not directly related to any other material property, it is most closely related to tensile strength. Corrosion, galling, and other surface defects reduce a material's resistance to fatigue more than can be associated with stress concentrations alone. Consequently, fatigue strength is usually determined experimentally. The fatigue properties of a material are based on testing specimens at different stress levels (S) and measuring the number of cycles to failure (N) to produce an S-N curve. Because real objects do not behave like test specimens, the design strength of a part is reduced based on:

- Whether the part is subject to flexural or axial loading
- The part size
- Surface condition (e.g., polished versus forged)
- Operating temperature
- Safety factor

FRETTING FATIGUE

Fretting fatigue is a special form of fatigue that is initiated by the small vibrations of parts that are pressed together. Vibrations of the parts cause surfaces to crack. These cracks produce stress concentrations that accelerate the fatigue of the parts. Fretting fatigue often occurs at pressed joints, which are not intended to move. It can be readily reduced in these cases by decreasing the clearance between the fastener and the through hole, or by using higher strength and correspondingly higher tightening torques. Impact loading may also result from oversize holes due to design or wear. Impact is undesirable because of the amplified stress it produces.

CREEP AND THERMAL RELAXATION

Creep describes the gradual stretching of a material under a load. Although it is most common in unreinforced plastics, it happens occasionally in metals also. Creep is due to slippage of a material's grains along their boundaries. This generally is not a concern except with highly loaded plastics that require dimensional stability and materials subject to high temperature. *Thermal relaxation* describes creep that is accelerated by high temperature.

Both fatigue and creep occur at stresses *lower* than the yield strength of a material. This means, for example, a drive shaft can break under normal loads after many years of reliable service. Furthermore, there will be no indicators of incipient fatigue failure. Corrosion of any sort greatly decreases fatigue strength. In fact, a part that has had pitting corrosion machined off will have much higher fatigue strength, even with the loss of material, compared to the original part with the pitting.

Many objects are designed for an infinite number of cycles but this evaluation can only be done by an engineer and must be verified by testing. A preventive maintenance schedule of critical components must be offered to properly ensure their reliability.

> ## Designing for Fatigue
>
> Here are some tips for designing for fatigue:
>
> - Avoid stress cycles.
> - Avoid brittle materials.
> - Avoid stress concentrations.
> - Do not let stresses exceed 50% of the yield strength for steel and 40% for aluminum and copper.
> - Look at the material's *notch sensitivity*.
> - Avoid operating near the melting point of the the material.
>
> Other minor factors such as surface, temperature, and residual manufacturing stresses also contribute to fatigue. Generally, rougher surfaces or those that have been heat treated have reduced fatigue life. One interesting aspect of surface treatment on strength is seen with the ancient art of hammering (cold forming) a sword surface to impart permanent compressive stresses that reduce the stress cycles that lead to fatigue failure.

Buckling

Like fatigue, buckling is a failure mechanism that occurs at loads well below the material design strength. The best way to observe the buckling phenomenon is to stand carefully on an aluminum can, which should support your weight. With your full weight on the can, have someone else use two rulers to bend the sides of the can in slightly and the can will buckle and crush rapidly.

Buckling is the collapse of a material under compression when the object bends and folds at loads that it would normally support (as we saw with the can). Buckling is a principle concern in axially loaded beams and thin-walled cylinders and is caused by the large increase in stress when a part is shifted slightly off center. Depending on the design of the part, this will make the part unstable and cause it to collapse. Buckling calculations can become quite complicated but they are a function of the slenderness ratio of a part. Long, slender objects will buckle under compressive loading, while short, stout ones will not. Figure 3-13 shows the types of compression failures associated with slenderness.

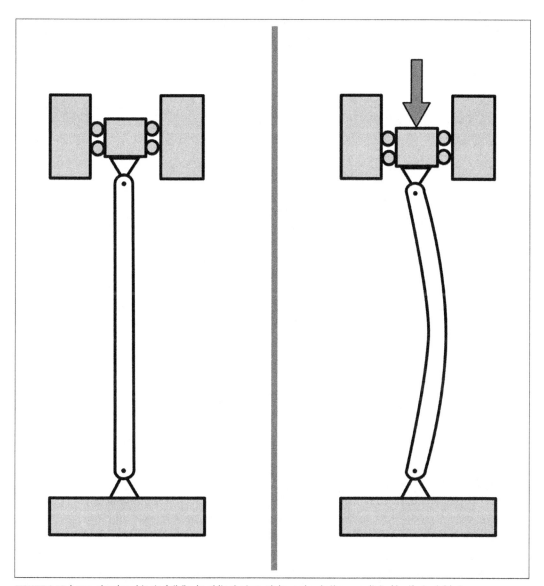

FIGURE 3-13. Long, slender objects fail (by buckling) at much lower loads than predicted by their yield stress

A Summary of Loads

These are common loads:

- Tension
- Compression
- Shear
- Flexure
- Torsion

These are some "weird" loads:

- Stress concentrations (amplifies loads)
- Fatigue (weakens material over stress cycles)
- Impact (amplifies loads)
- Buckling (fails below expected expected compressive strength)
- Thermal expansion (induces stress or changes dimensions)
- Corrosion (reduces area)

 Don't forget unexpected loads such as those associated with nature, shipping, and incorrect use!

Thermal Expansion

Hot stuff expands, cold stuff contracts. If an object is not allowed to move freely, it produces high stress as it grows into an adjoining object. Heat increases atomic vibration and makes the atoms act as if they are bigger. Therefore, thermal expansion and temperature are directly related. The amount of expansion varies by material and is measured by the thermal expansion coefficient. Brass has a thermal expansion that is 50% more than steel. Aluminum thermally expands at twice the rate of steel, as shown in Table 3-2.

TABLE 3-2. Coefficients of thermal expansion at 70°F (21°C). This data indicates the amount of expansion a 1" or 1mm material experiences for every degree of temperature change.

Material	Expansion coefficient (in./in./°F)	Expansion coefficient (mm/mm/°C)
Steel	0.000006	0.000011
Brass	0.000009	0.000017
Aluminum	0.000012	0.000022
Cast iron	0.000005	0.000009
Polyethylene[a] (high density)	0.000066	0.000119
Polycarbonate[a]	0.000040	0.000072
Nylon[a]	0.000056	0.000101
Plate glass	0.000005	0.000009

[a] Expansion coefficients for polyethylene, polycarbonate, and nylon (polyamide) vary widely based on specific composition

Thermal expansion or contraction is calculated by:

$$\Delta L = L_1 \, \alpha \, \Delta T$$

where:

ΔL = change in length after heating or cooling
L_1 = initial length
α = coefficient of thermal expansion
ΔT = temperature change

If two materials expand into each other, either because they have different thermal expansion coefficients or they are at widely different temperatures, they develop stresses directly related to the strain they produce. That is, the strain produces stress. If a column or pipe is anchored at both ends and is axially constrained, the compressive stress generated by thermal expansion becomes high. Plastics are viscoelastic, meaning they will flow like a liquid. Therefore, they can reduce the stresses caused by thermal expansion to some degree by fluid flow.

The compressive stress generated by thermal expansion is calculated by:

$$\sigma = \alpha \, E \, \Delta T$$

where:

σ = compressive stress
α = coefficient of thermal expansion
E = modulus of elasticity
ΔT = temperature change

We are designers, not vibrating molecule specialists, so why should we care about this? Thermal expansion can break things and change their shape. While you might think your product lives a benign life on a store shelf and in someone's home, when it is shipped in the belly of an airplane or on a giant containership plying the north Pacific, it is exposed to freezing temperatures. When it is stored in an attic or transported by truck, it can get very hot.

The Design of a Tree

Much of what has been described can be highlighted in the design of a tree:

Stress concentration
    The branches sprout off with a blend fillet to reduce stress concentration.

Flexural strength
    The branches get slimmer the further away from the trunk (as does the trunk the further from ground level). Because the moment arm is longer where the branches join, the cross section must be larger.

Moment of inertia
    Because the branches are round (like our bones), the branches are torsionally rigid and have the same moment of inertia in all planes. Therefore the branches behave the same due to the downward pull of snow or the sideways push of wind.

Impact
    The branches are not overly rigid so they sway in response to wind gusts, therefore reducing the speed at which the load can be applied.

Buckling
    The trunks don't get too long and skinny for the compressive load they might bear.

The force of thermal expansion creates high stresses when two materials with dissimilar thermal expansions (e.g., plastic and metal or aluminum and steel) are entrapped. The stresses can also be produced when a hot object is completely constrained by another object, which is why bridges have thermal expansion joints. Thermal expansion is a special problem for big items or those that experience large temperature changes because the effect is multiplied. Therefore, such things as piping systems and engine exhaust need to be thermally accommodated by including flexible joints or bends to the system's design.

Failure Modes

You have built a prototype and it broke. Now what? Well, there is a lot of good news in the broken parts you are looking at. They contain a wealth of data to help you see what went wrong. Even if your device doesn't break, it is still considered "failed" if it stops working the way it should. This section describes different ways devices stop working, from breakage to changing their shape and jamming.

ELASTIC DEFORMATION

This failure mode is caused by a load on a part high enough to distort the part to the point it will stop functioning. In other words, nothing breaks but the part shape changes so much under its load that it jams or locks up in some fashion. The load can be caused by an external force or thermal expansion. In the case of elastic deformation failure, the load distorts the material within its elastic range. That is, the stress induced by the load is low enough that the material recovers completely when the load is removed. If the stress is higher than the yield strength of the material, it will stretch and never recover. If the stress exceeds the ultimate strength of the material, the material will break.

YIELDING

Yielding failure is similar to elastic deformation in that the part does not actually break. Unlike elastic deformation, the part will not return to its unstressed dimensions when the load is removed. Yielding occurs when the applied force exceeds a material's yield strength. A part that has yielded is stretched, distorted, and is close to breaking. Ductile materials, such as steel, stretch a great deal before a ductile rupture occurs. Brittle materials, such as glass and fiberglass fiber, do not yield before a brittle fracture occurs. Consequently, ductile materials give warning signs before they break while brittle ones do not.

DUCTILE RUPTURE

Under a continually increasing load, a ductile material will elastically deform, yield, and then finally rupture. A ductile rupture is caused by the slow growth of internal cracks under a load. The cracks grow together until the part breaks. The fracture face will usually be rough and have a "torn out" appearance. This is due to the load shearing up and down between the grains.

BRITTLE FRACTURE

Unlike a ductile material, a brittle material has little yield and can fail without warning. A brittle fracture is caused by the rapid separation of interatomic bonds. The internal voids and cracks that are always present in materials initiate this fracture. In brittle materials, the stress concentrations produced by these features cannot be distributed. The fracture face of a brittle fracture will usually appear as a flat, crystalline surface and may show chevron-shaped beachmarks that point to the failure origin.

A brittle fracture can occur in normally ductile materials at low temperatures (below the *transition* or *glass temperature*) or when they have been subjected to *stress corrosion cracking*. When the temperature of a metal is below this transition temperature, a crack can travel faster than the metal can deform. Because deformation normally absorbs tremendous amounts of energy, a low-energy crack that does not produce deformation can propagate readily.

FATIGUE

Fatigue is a failure mode caused by cyclic action of a part resulting in failure at a below-normal stress. This is by far the most common cause of mechanical failure. A fatigue fracture will show two distinctive surface appearances. One of these areas exhibits curved ridges or beachmarks that were produced as cracking slowly progressed through the object. The other surface will usually be rougher and show no distinct pattern. This rougher surface is created when the object is so weakened by the fatigue that it ruptures quickly.

IMPACT

Impact is a very rapidly applied load such as a hammer striking a nail. Stress waves, produced during impact, actually stack up on each other, amplifying the stress. Figure 3-14 shows an example of impact power.

FIGURE 3-14. Jackhammers amplify force by impact

BUCKLING

Buckling occurs in parts that have a very low thickness-to-length ratio. Buckling is a common failure mode in thin-walled tubing and unsupported reinforcement fibers. It will occur at stresses much lower than the compressive strength of the material.

THERMAL SHOCK

This failure is well described by its name. It is caused by such large differences in thermal expansion within a material that it fractures or yields.

WEAR

Wear is the result of one object contacting another, leading to the removal of material. Wear is not necessarily due to the plowing of the object, specifically referred to as abrasive wear, but can also be due to adhesion of two mating materials. This adhesion causes surface cracks that grow and eventually cause small flakes of material to break away. This type of wear is called *galling* and if the wear continues, the galling will lead to seizure of the two wearing parts.

Cavitation is an erosive wear caused by the formation and implosion of bubbles. The shape of a flowing object, such as a propeller, can produce a pressure sufficiently low to boil a fluid. When the vapor recondenses, it implodes and wears material away.

BRINELLING

Brinelling fractures occur when two curved surfaces are excessively loaded so yielding occurs over the small contact area. The most common example of brinelling is observed in ball bearings that have had an excessive radial load applied to them. This loading forces the balls against the inner part of the bearing (race) with sufficient force to brinell the surface and leave small dents in the race.

SPALLING

Spalling is caused by cracks developing between the grains or crystals of a material. The cracks grow until particles flake off. Spalling often results from brinelling, corrosion, or high temperatures.

Failure Modes and Analysis (Learning from Failure)

Diagnosing the root cause of failures is difficult. The forensic techniques used can range from visual observation to chemical analysis. However, a keen observational skill can often deduce failure mechanisms by using a combination of direct observation of the fractured part under investigation and the conditions associated with the failure. These associated conditions can include types of loading (tension, compression, shear, flexural, or torsional), corrosion, surface marks (suggesting misuse), and other clues in the conditions associated with the failure.

A part has "failed" when it does not perform its function. This is true even when the part is not broken. For example, a hinge has failed if it heats up and expands during operation causing a linkage to lock up. Specifically, the hinge was subject to a temperature-induced elastic deformation failure. This semantic review is necessary in order to introduce the precise terminology of failure analysis. The following paragraphs describe common failure modes. Failure is often the result of several concurrent effects. These combined effects can lead to failures at stresses well below the maximum strength of a material, as is the case with fatigue and stress corrosion cracking.

In identifying the cause of failure, the methodology of *root cause failure analysis* should be followed. This basically involves the following steps:

1. Identify the *primary failure*—that is, the first component that failed. This failure may have led to secondary failures that are irrelevant to finding the root cause of failure.

2. Learn about the circumstances around the failure. What happened to make the part fail? How were loads applied? How old is the part?

3. If the primary failure was the breakage of a component, inspect the surface where the break occurred, referred to as the *fracture face*. This face will give an indication of the ductility of the material as well as whether the failure was due to an overload or fatigue.

4. Metallurgical or chemical analysis can reveal whether the material is defective. Detached, irregular cracks in metal parts are usually caused by a material defect.

5. Consider all the information gathered and determine the root cause. For example, if a part has been in service longer than recommended by the manufacturer and subsequently failed due to fatigue, poor maintenance would be considered the root cause of the fatigue failure.

Noise and Vibration

Noise and vibration are important issues in design. They impact the health and experience of a user. Both noise and vibration are mechanical phenomena, whether they be oscillating air molecules or solid material. While noise can be harmful, its gentle cousins of music and *sound* can be reassuring, comforting, and an essential part of a user experience. Vibration can help us understand how something is responding, whether it be damaged bearings or an intentional haptic feedback system. However, vibration can also produce fatigue failure and other damage to mechanical systems.

NOISE

Sound is one of the three senses, along with visual and tactile, to which the designer must appeal. (Yes, you could say four—who can deny the appeal of the smell of leather or jasmine?) Sound communicates. The zipping sound of a zipper tells us it is working, the snapping sound of a clip tells us it is engaged. The solid sound of a car door closing suggests quality. Sound is part of the designer's palette that can inform and appeal to the user. Sound can often move from satisfying and reassuring to obnoxious noise. We get the squeaky shoe and the screaming vacuum cleaner. Unfortunately for the consumer, these sounds don't become obvious until after purchase. Moreover, sounds are often more offensive to those who must listen as opposed to those who create the sound. It is more fun to hammer than to listen to a hammer. Noise is defined as an erratic order of various frequencies. Besides being bothersome, noise can have a physiological impact by producing hearing loss and physical stress.

Sound differences are measured in the logarithmic scale of decibels (dB). From the ticking of a wrist watch to the roar of a shotgun, the differences between the softest and loudest sound we can hear before eardrum rupture is a factor of many trillions.

However, it is not merely the loudness of the noise that is offensive. High frequencies are the most obnoxious and dangerous to humans. These obnoxious frequencies are captured in the weighted decibel scale (the standard unit for measuring sound). Weighted scales take the decibel readings in the frequency range of human hearing and heavily multiply the higher frequencies. The most popular weighted scale is the dBA scale. With this weighted scale, a sound that has a loud 10,000 Hz component would record a much higher dBA reading than does one with a loud 500 Hz component.

Acoustic resonance is the term used to describe the oscillation of an object at its natural frequency. When a tuning fork is struck, it will vibrate at a frequency dictated by its rigidity (moment of inertia) and material composition. Resonance is an important concept because the resonant frequency of parts near vibration sources should be designed to fall either at a very low frequency or at one well above the level of human hearing.

The best way to reduce noise is to locate its source and eliminate it at that point. For example, "engines" do not produce noise; it is certain components or features of the engine that produce the noise. Some of the root causes of engine noise are flowing inlet air and exhaust air, cylinder liners, gears, and shafts. These sources can be located by acoustic intensity measurements. Acoustic intensity measurements are made with a series of microphones and processing equipment that locate the exact position of a noise source.

The approach to noise abatement, excluding air flow noises for the moment, is to first identify the loudest noise source. The offending part may be temporarily replaced with a quieter material to verify it is the noise generator. Once verified, the object will be redesigned to reduce its noise. Often this involves changing its rigidity to move its natural frequency. This locating process is continued until the noise production is reduced as much as possible. After this is done, acoustic panels, mufflers and other more traditional noise abatement techniques can be used.

Basically, acoustic panels work by using the noise energy to vibrate foam. The soft foam does not transmit the noise well and thereby expends the noise energy and reduces further transmission. Acoustic panels are often made with a dense sheathing attached to the foam. This "tuned" system allows greater noise energy dissipation. Acoustic panels are often configured with a couple of stacked sheathed cores with the innermost sheathing made from a sheet of perforated metal.

VIBRATION

Vibration can be reduced by changing the rigidity or by vibration isolation. Rotating equipment is often mounted on flexible mounts that reduce the transmission of vibrations to the equipment.

Torsional vibrations are oscillatory movements arising in rotating equipment and can manifest themselves as gear chatter. A rotating shaft's torsional rigidity, length, and the moment of inertia dictate the frequency at which the shaft oscillates back and forth. However, the shaft's resonant frequency can be moved above the operating speed by stiffening the shaft or using a flexible coupling. Viscous dampers are sometimes mounted to the end of shafts to absorb minor torsional vibrations.

Closing Thoughts

Now for the juggling! You know the careful shaping of an object's cross section lets you use the minimum amount of material and avoiding sharp corners reduces stress concentration. You know long, skinny shapes buckle, high loading cycles produce fatigue, impact amplifies stresses, corrosion can be hidden, and materials expand when they are heated. All of these factors go into keeping the loads under the yield stress of the materials.

If you were designing a water piping system, you would need to worry about many more things than simply its ability to move water from one place to another. You would think about what happens to the pipe when the water is turned off quickly (impact), what happens when it heats up (expands), and what happens at the elbows (stress concentration). You would also be thinking about the flexing of the pipe, the axial loading of the pipe supports, and the corrosion of fittings.

If you were designing a fancy new bicycle wheel, you would need to think about the load of the weighted bike traveling through the wheel, the impact of potholes, and the corrosion produced by water and road salt (not to mention what that will do to the bearings!) You would also think about how a rider might get their foot dangerously tangled in your wheel design and the side loads the wheel encounters if it tips over sideways. You would need to consider the pressure of the tire on the wheel and what the minimum cross section should look like. You know any cracking will probably originate from the sharp corners in your design and may not show up for a long time.

When you design a frame to support your new invention of a dog trampoline, you are thinking about the flexural stresses in the frame, their propensity to buckle, fatigue strength, corrosion, and vibration (maybe the frame resonates at the dog's favorite jumping rate!). Welcome to the world of worry and good design.

CHAPTER 4

Materials

ONE OF THE MOST DYNAMIC AREAS OF DEVELOPMENT RELATED TO DESIGN IS IN MATERIAL SCIence. Design is visually enhanced by material selection. The allure of gold seems to be driven into our DNA while the textured grit of iron seems to suggest permanence. We love the changing patina of copper, the warmth of brass, the shine of aluminum and silver, the technical appeal of carbon fiber, the understated strength of titanium, and the invitation to touch by leather, wood, and texturized plastic.

Even though this is not a material science book, we have to give materials some mention because materials can save the day for a design in unexpected ways. I have had designs saved by such things as beryllium copper that will never spark, polyethylene that could be purchased in extremely long lengths, fluorinated greases that will not dissolve in fuel, and exotic coatings that provided high temperature service and wear resistance.

We have already considered material strength and stiffness, which include yield and ultimate strength as well as its response to stress (modulus of elasticity). However, there are many other important material attributes. For example, a material's hardness is a characteristic that has practical importance in determining a part's strength and wear characteristics. Hardness is defined as the resistance to penetration and is directly related to material strength. Metal hardness is usually changed by various heat-treating processes that modify the grain structure.

Other important material properties include density, dimensional stability, toughness, impact resistance, temperature weakening, water absorption, ultraviolet light resistance, flammability, thermal resistance, and a range of electrical, magnetic, and optical properties.

These are some common mechanical properties of materials:

Density
 Mass per volume.

Hardness
 Resistance to deformation. It relates to strength among other things; however, often surface hardness is made different than the material hardness through heat treating or other surface treatments.

Impact resistance
 Ability to resist a rapidly applied load.

Modulus of elasticity (Young's modulus)
 Relationship between amount of stress and amount of stretch (stress/strain). This is the slope of the stress/strain curve.

Toughness
 Amount of energy a material can absorb before breaking. This is the area under the stress/strain curve.

Ultimate strength
 Stress that causes material to break.

Yield strength
 Stress that starts to permanently stretch material.

These properties are most closely considered in plastics:

Dimensional stability
 Resistance to dimensional change due to temperature and humidity.

Glass transition temperature
 Temperature above which a hard material becomes soft. Note: metals have a ductile to brittle transition temperature.

Water absorption
 The rate at which water is taken into a material.

Ultraviolet resistance
 Resistance to degradation by ultraviolet radiation.

Flammability
 Propensity to burn.

Classification of materials:

Homogenous
A material that has the same elastic properties at all points.

Isotropic
A material having the same elastic properties in all directions at any point in the material.

Anisotropic
A material that is not isotropic with varying properties depending on direction.

Orthotropic
A material that has three mutually perpendicular planes of elastic symmetry.

 While this list does not include electrical, magnetic, and optical properties, they are important too!

Selecting a material is a balance between performance, economy, and fabrication. But it is difficult to optimize on material selections. Sometimes favorites evolve based on experience and the following is an overview of some key materials—but remember the world of materials changes quickly! Where material properties are given, they provide a nominal value that is best used for comparison because material designations relate to chemistry and often do not provide the details related to such factors as mechanical properties, heat treatment, or quality.

Materials basically fall into one of the following categories:

- Metals
- Polymers
- Elastomers
- Ceramics
- Glasses
- Natural materials (such as wood and stone)

Characteristics of Metals

Strong, reliable, and the namesake for great Ages of history, metals produced the first industrial revolutions. Like wood, metals present appealing lusters ranging from the rich warmth of polished gold to the crusty slag of wrought iron that belies the maelstrom of fire from which it originated.

STEEL

Steel is cheap and can generally give the highest strength-to-dollar ratio of any material. Beyond the pedestrian applications of steel, some alloys achieve exotic successes and can be sorted into the following categories: high strength, high toughness, high temperature, high hardness, and high corrosion resistance. Many nonferrous materials excel in these areas but usually at a higher cost, such as titanium for high temperature or gold for high corrosion resistance!

Carbon is the key element in determining steel characteristics. A small increase in carbon content greatly increases strength and hardness but reduces ductility. With low ductility, cold forming processes like stamping become more difficult. Cast iron is generally defined as "steel" containing more than 2% carbon. Cast iron is difficult to machine and bend, therefore shapes are normally produced by casting. The bulk of steel consumption falls in the range of low carbon steel. These steels contain less than 0.3% carbon and are interchangeably referred to as "low carbon" or "mild" steel. Steels with 0.3–0.6% are referred to as medium carbon steels and those with more than 0.6 percent are called high carbon steels.

Carbon content is so important in determining steel characteristics that it is embedded in the standard nomenclature for steel. For example, a popular cold-rolled steel is referred to as 1020. This number derives from SAE (Society of Automotive Engineers), AISI (American Iron and Steel Institute), and ASTM (American Society for Testing and Materials) designations. The "20" in these designations indicates the steel has a carbon content of 0.20%. Note the Unified Numbering System (UNS) is used in North America and includes a letter to indicate alloy type (e.g., C for copper, S for Stainless, and a five-digit number sequence); however, the four-digit and three-digit designations are very commonly used and will be presented here. For example, AISI 1018 is equivalent to UNS G10180. International equivalence can readily be found for UNS or SAE/AISI designations. It is important to note these material designations do not completely specify a material because they do not include material properties, heat treatment, or quality.

Many additives are routinely added to steel to ensure quality by reducing or sequestering impurities. For example, to remove crack-promoting sulfur, a small amount of manganese is added to combine with any residual sulfur. Silicon is also added to prevent oxygen from dissolving into the steel grains and making them brittle. Many other elements such as aluminum, cobalt, copper, lead, manganese, molybdenum, silicon, and tungsten are added to steel to develop specific properties. These steels are often referred to as "high alloy" steels. Tool steels are alloys of cobalt, molybdenum, tungsten, and vanadium specifically blended for cutting operations. Chromium alone or chromium and nickel produce alloys known as "stainless" steels. Using the four-digit system, the first number of some common steel alloys are 1 = carbon, 4 = molybdenum, 5 = chromium, 6 = chrome vanadium, 8 = nickel chromium molybdenum, and 9 = silicon manganese.

The second number in the designator indicates whether there are other significant alloying elements. So 1018 steel has no other alloying elements. Some other common alloys such as 1113 and 1141 have sulfur added to them to improve machinability. Letters are sometimes added in the middle of the designation such as B for boron or H for hardenability. For example, 11B41 is 1141 with boron added to increase the hardness. As mentioned before, the carbon content is given in the last two numbers; however, the actual carbon content can vary slightly from this percentage.

COMMON STEEL ALLOYS

Common steel alloys include 1010, 1018, and 1020, which can be easily formed and welded. 4130 is commonly used in aircraft because it can be welded and hardened. 4140, 4340, and 6150 have high impact resistance and can be deep hardened.

Cold-formed steel, in which the steel is deformed at "cold" temperatures, creates a permanent increase in the hardness and strength of the steel. In order for these improvements to be sustained, the temperature must be below a certain range, because the structural changes are eliminated by higher temperatures.

Hot-rolled steel is heated and easier to form into a shape. Using the common three-letter designation (I know this gets confusing), we must comment on the ubiquitous ASTM A36 HR. It is cheap and strong but contains impurities, resulting in problems during machining. A36 has a yield strength of 36,000 psi (248 MPa), so it is a good material when your design will live a peaceful existence and is not subject to denting or other misuse.

STAINLESS STEEL

Commercialized stainless steel developed quickly from research in the early 1900s. This early research was rooted in metallurgical studies of chromium in iron that went back as far as 1821. Stainless steels have provided the same reliable characteristics as steel without the problem of rapid corrosion. The term "stainless steel" was actually a trade name for the new chromium-nickel steel alloy. The easy-to-remember trade name has stuck. The austenitic stainless (300 series) steels contain chromium and nickel that combine to give the alloy excellent corrosion resistance. This type of stainless steel is very weakly attracted by magnets (referred to as *paramagnetic*)—notably different than the eagerly magnetized carbon steels.

The austenitic stainless steel cannot be heat treated, therefore it cannot be hardened like the other stainless steels. For example, low-chromium versions of ferritic stainless steels can be heat treated and are commonly used in cutlery. Many of the ferritic stainless steels are strongly attracted by magnets (referred to as *ferromagnetic*). This magnetic attraction challenges the time-honored method of differentiating stainless and carbon steels. Most stainless steel fasteners are made from austenitic stainless steel so the easily observed difference in magnetic attractions is usually an accurate method of differentiating between stainless and carbon steel. Stainless steels have a 30% higher thermal expansion coefficient and one-third the thermal conductivity of carbon steels.

Common commercial grades of stainless steel include the following SAE/AISI Types:

303

Contains 0.15% carbon, 17%–19% chromium, 8%–10% nickel, 2% manganese, and 0.15% sulfur. The sulfur is added to allow for easier machining. This grade is often used for fittings and other objects that require fine machining.

304

This alloy contains a maximum of 0.08% carbon, 18%–20% chromium, 8%–10% nickel, and 2% manganese.

316

Has higher corrosion resistance than 304, especially to acids. It contains a maximum of 0.08% carbon, 16%–18% chromium, 10%–14% nickel, 2% manganese, and 3% molybdenum.

Alloy modifications of these common stainless steels are often made and are noted by the addition of a letter. Most interesting are AISI 304L and AISI 316L. These steels have very low carbon content with a maximum of 0.03% and are used for high temperature applications.

ALUMINUM

Aluminum is a popular fabrication material, well known for its lightness and renowned for its ability to develop a highly polished luster. Like steel, there are numerous alloys of aluminum that are intended to produce a wide range of properties.

Alloys are designated by a variety of nomenclature; broad ranges of alloys can be summarized by the following series of numbers:

1000

Nearly pure aluminum, they are commonly used in electrical and chemical applications.

2000

Copper is the principal alloying element. Copper improves the strength of the alloy. However, it does not have good corrosion resistance and is subject to intergranular corrosion. 2024 is commonly used for aircraft components.

3000

Manganese is the principal alloying element. 3003 is a commonly used alloy because it combines strength, low cost, and formability. Formability is the ease in which a material can be bent, stamped, or otherwise deformed. The formability characteristic reflects a wide difference between the yield and ultimate strength.

4000

Silicon is the principal alloying element. The silicon lowers the melting point, so it is commonly used for welding wire. Some of these alloys can be treated so they turn a dark gray color. These alloys are popular for architectural applications.

5000

Magnesium is the principal alloying element. Magnesium is more effective than manganese in improving the alloy strength. These alloys have good resistance to corrosion.

6000

Magnesium and silicon are the principal alloying elements. This alloy combines the attractive characteristics of strength, corrosion resistance, and formability. This alloy can also be readily heat treated.

7000

Zinc is the principal alloying element. High strengths can be achieved with these alloys and they are used in the aviation industry for highly loaded components.

Brass

Brass is corrosion resistant, and easy to mold and machine. Brass also has well-appreciated aesthetic attributes. Brass is an alloy of copper and zinc that offers an unusual set of qualities. Not only does this combination of elements increase strength, which typically occurs with an alloy, but it also increases the alloy's ductility. Usually ductility decreases when a high-strength alloy is produced.

The major families of brasses are the copper-tin-zinc alloys, manganese-bronze alloys, leaded manganese-bronze alloys, and copper-zinc-silicon alloys (including silicon bronze).

Table 4-1 shows the composition of common brass alloys.

TABLE 4-1. Composition of common brass alloys[a]

Alloy	Copper	Zinc	Tin	Lead
Gilding brass	95	5		
Commercial bronze	90	10		
Red brass	85	15		
Cartridge brass	70	30		
Yellow brass	65	35		
Free machining brass	61	36		3
Naval brass	60	39	1	

[a] Percentage, actual composition, and trace elements vary.

BRONZE

Bronze was traditionally an alloy of copper and tin. However, the true definition of a bronze is an alloy of copper and any material besides zinc or nickel. The copper-tin alloys are very similar to brass. However, alloys of copper with aluminum, beryllium, and silicon produce some special characteristics. For example, aluminum bronzes can be heat treated to obtain relatively high strength. Copper beryllium is also heat treatable, very strong, resistant to sparking, and fatigue resistant.

The major families of bronzes are the copper-tin alloys (tin bronzes): copper-tin-lead alloys (leaded tin bronzes); copper-tin-nickel alloys (nickel is not a major alloying element); and copper-aluminum alloys (aluminum bronzes). Manganese bronze includes zinc as a major alloying element and therefore is a brass, not a true bronze. Copper-nickel alloys are actually a separate category of copper alloys that are very resistant to corrosion. Copper-nickel-zinc alloys (nickel silvers) fall into their own special category.

Table 4-2 shows typical compositions of bronze alloys.

TABLE 4-2. Composition of common brass alloys[a]

Alloy	Copper	Tin	Zinc	Other major elements
Beryllium bronze	98			2 Beryllium
Silicon bronze	96			4 Silicon
Tin bronze	88	10	2	

[a] Percentage, actual composition, and trace elements vary.

NICKEL

Nickel is usually mixed with copper to make some remarkable alloys. They are usually difficult to machine but have remarkable corrosion and high-temperature characteristics. Monel is a common alloy comprised of two-thirds nickel and one-third copper.

Nickel is often the contributor to alloys with amazing characteristics ranging from the Nilvar alloy (nickel and iron) with its tiny 0.000001 in./in./°F (0.000002 mm/mm/°C) thermal expansion to the Inconels (nickel-chromium-iron) that have been used for such high-temperature, high-strength applications as the fuselage of the X15 airplane.

MAGNESIUM

Magnesium is two-thirds the density of aluminum, which is its most significant advantage. It is used when weight is important but titanium is too expensive or difficult to process. Magnesium is often used in die casting to obtain exquisite details. Magnesium chips are very flammable and are used in fireworks and fire starters. Magnesium is often alloyed with other metals such as aluminum and zinc.

TITANIUM

Titanium has the best strength-to-weight ratio of any metal and is best known for its feats in aerospace applications. It is also corrosion resistant, especially to seawater and acids. Commercial alloys have yield strengths similar to the austenitic stainless steels. However, exotic alloys can have yield strengths of 200,000 psi (1,378 MPa).

ZINC

Zinc is commonly used for galvanizing, which is a thin, protective layer bonded to steel that provides corrosion resistance (see the section "Corrosion Behavior of Materials" on page 82). Zinc is also used to make high-definition die castings. Zinc has a low melting point of 787°F (419°C) and boils at 1665°F (907°C), so it cannot be welded at a high temperature such as produced with arc welding. Zinc is often alloyed with other metals in small quantities but processes such as casting or brazing may reduce the zinc concentration because of its low melting temperature.

Table 4-3 shows the mechanical strengths of common metal alloys.

TABLE 4-3. Mechanical strength of common metal alloys[a]

Material	Typical alloy	Yield Strength, Tensile x 1,000 psi (MPa)	Ultimate Strength, Tensile x 1,000 psi (MPa)
Magnesium	SAE AM265C	11 (76)	27 (186)
Aluminum	ANSI 3003-H14	21 (144)	22 (152)
Aluminum	ANSI 6061-T6	40 (275)	45 (310)
Aluminum	ANSI 7075-T6	73 (503)	83 (572)
Nickel-copper	Monel 400	30 (207)	79 (544)
Titanium	RS 140	140 (965)	150 (1,033)
Naval brass	UNS C46200	53 (365)	75 (517)
Malleable iron	SAE M3210	32 (220)	50 (345)
Steel	ASTM A36 HR	36 (248)	58 (400)

Material	Typical alloy	Yield Strength, Tensile x 1,000 psi (MPa)	Ultimate Strength, Tensile x 1,000 psi (MPa)
Steel	AISI 1018	54 (370)	64 (440)
Steel	AISI 1040	42 (289)	76 (524)
Steel	AISI 304	35 (241)	85 (586)
Steel	AISI 316	35 (241)	85 (586)

[a] Properties vary widely depending on processing and heat treatment. Properties for steels are based on hot rolling. Cold-drawn versions of the same grade are work hardened during the process and have higher strength values. The aluminum strength values are based on their specified temper. The naval brass strength value is based on hardened rod.

Surface Wear

When solids slide against each other, their surfaces are degraded by abrasive, surface fatigue, adhesive, and tribochemical wear. Abrasive wear is caused by plowing between two materials in which one or both is then plastically deformed and fails. Surface fatigue wear, caused by cyclic stresses, initiates and propagates cracks. These cracks grow large enough to cause the surface material to flake. The tribochemical wear mechanism is a combination of mechanical and thermal processes at the contact interface, which increases the corrosiveness of the surface.

When solid materials move against each other, friction is produced between particles where the peaks (asperities) and valleys of the particle surfaces interface. This contact area is much smaller than would be provided by an apparent contact area of a perfectly smooth surface. Generally, the smallest of these asperity junctions contact one another under highly localized pressure and therefore deform plastically. With further load, however, the larger asperities engage each other and the contact area increases. These asperities deform elastically because of the large contact area. With many metals and brittle materials the mechanism of plastic deformation is an anisotropic "slip" in which planes of atoms slip over each other. As the load increases, a critical shear stress is achieved causing plastic deformation within the zone of elastic deformation. Clearly, the task of quantifying asperities is difficult because of the small contact asperity junctions and the subsequent alacrity in which they deform. Moreover, the contact of two curved surfaces, such as a ball bearing and raceway, present triaxial stresses that causes spalling.

Most metals, such as aluminum on aluminum or aluminum on copper, don't like to slide against each other. However, copper alloys slide against steel and cast iron with ease. Gray iron slides against these metals because of the lubricity provided by the free graphite in the cast iron. Steel can slide on steel if both faces are hardened. Lubricants, such as oil, Teflon, graphite, and molybdenum disulfide, are usually used to allow metals to slide.

Characteristics of Ceramics

Ceramics are derived from some of the earth's most abundant minerals, including bauxite, clay, silica, feldspar, and talc. Ceramics cover a broad range of applications, from Chinese porcelain to heat shields on space craft. They are normally crystalline, hard, strong, brittle, and chemically impervious. Ceramics are well regarded for their excellent high-temperature performance—they can operate at temperatures up to 3,000°F (1,650°C). Their hardness allows them to be used effectively in cutting tools and abrasives. They enjoy an ancient history as pottery and construction materials.

Characteristics of Plastics

Plastics have revolutionized designs because of their remarkable ability to do wonderful things such as being slippery, strong, colorful, noncorrosive, and nonconducting. Currently, the main areas of weakness for plastics are with very low- or high-temperature applications and in applications requiring high surface hardness.

Plastics are usually based on long, repeating molecular arrangements called *mers*. This term is the root of *polymers*. Plastics have the same structure as all organic substances. They are composed of long chains of carbon atoms with hydrogen atoms branching off each carbon atom. Additional atoms are attached to these mers to give them their unique characteristics. For example, polyvinyl chloride (PVC) has a chloride atom attached to its mer. The molecular structure of organic molecules can be seen in everyday life. Their hydrocarbon structure can easily be altered by heat. When wood, bread, PVC pipe, or other organic structures are overheated, they become charred. That is, the hydrogen atoms linked to the outside of the hydrocarbon chain are driven free, leaving only the dark-colored carbon backbone.

The *glass transition temperature* of plastics is an important parameter for consideration. The glass transition temperature (or simply *glass temperature*) is the temperature below which the molecules cannot rearrange themselves when a load is applied. Therefore, at temperatures below the glass temperature, the plastic becomes rigid and brittle. Hard plastics, like the polystyrene used in plastic models, are used above this temperature while elastomers, like rubber, are used below this temperature.

Polymers have an inherent spring-like tendency because the bonds that hold the atoms together are at an angle. The amount of "springiness" is usually controlled by the amount of interconnection between the mers. As the mers become more interlocked, they transition from *elastomers* to *thermoplastics* and finally to *thermosets*.

Polymers can develop into a three-dimensional structure during initial fabrication. The three-dimensional structure makes the movement of molecules difficult when heated; therefore, once formed they do not soften at high temperatures. Consequently, they cannot be molded. This type of polymer is called a thermoset. Unlike thermosets, thermoplastic bonds disappear at high temperatures and thermoplastics soften at high temperatures. Therefore, these polymers can be easily molded and will only hold their shape when cooled.

Natural rubber, silicone, nitrile, fluoroelastomer, and other flexible materials are called elastomers. The glass transition temperature is especially important in elastomers because their desired quality of flexibility will completely disappear below that value. In fact, a rubber band that has been frozen in liquid nitrogen can easily be snapped like a piece of glass.

The main problem with most plastics is their poor performance at high temperature and their flammability and propensity to generate a lot of putrid smoke. Plastics are also prone to weakening from ultraviolet light, water absorption, dimensional instability, creep, and poor impact resistance

New versions of plastics are entering the market every day. The old standby "commodity" plastics (polyethylene, polypropylene, and polystyrene) and "engineering" plastics such as ABS (acrylonitrile butadiene styrene) and nylon have been tremendously enhanced over the years. Newer entries with complicated chemical designations and equally odd trade names have added even further to plastic's market penetration. Many plastics can be modified to provide such enhancements as ultraviolet light resistance and internal lubrication. Moreover, the strength of almost any plastic can be increased by adding small, randomly oriented fiber reinforcements.

Table 4-4 shows the mechanical strengths of common plastics.

TABLE 4-4. Mechanical strength of plastics

Material	Tensile yield strength:	x 1,000 psi	MPa
Polyethylene, high density		4	27
ABS		6.5	44
Polycarbonate		9	62
Polyester		10	69

Characteristics of Foams

Plastics are used to make foams such as Styrofoam. Foams are comprised of air with a thin matrix of solid material, typically urethane, polyester, and PVC. With all the air in the matrix, they are not very strong, but they are light and good insulators. Foams can be used alone for a variety of purposes, from thermal insulation to packing. They can also be used as the coring in sandwich construction.

Closed cell foams, as opposed to open cell, have sealed bubbles and therefore don't absorb moisture. These are commonly used in exercise mats, sound absorption, and thermal insulators. Anti-static, conductive, and static foams are used for electronics packaging as well as sound proofing.

Some commodity foams include polyurethane, which is a cheap packing foam and has a tensile strength of less than 7.5 psi (52 kPa). High-density foam is used for furniture and has a tensile strength of less than 15 psi (104 kPa). Higher-grade foams are also used in furniture such as high-resilience foams. Viscoelastic (memory) foams are used in mattresses and medical supports.

Characteristics of Wood

Wood is a cellular material. Specialized cells provide structural support, and transmit sap or store food in the living tree. Wood cells are composed of cellulose, lignin, extractives, and ash-forming minerals. Wood is composed of 70% cellulose. Lignin acts as the adhesive matrix for wood. The ash-forming minerals constitute less than 1% of the wood and contain the stored nutrients for the tree. The extractives are not part of the wood structure but produce the characteristics of color, odor, and decay resistance. They include tannins, starch, oils, resins, fats, and waxes.

A cross-section of a tree shows two different colored regions. The lighter, outside ring is *sapwood* and the darker inside core is *heartwood*. Sapwood is usually less than two inches thick (5 cm). However, with ash, hickory, maple, and some pines the sapwood ring may be six inches (15 cm) thick. The sapwood consists of living cells that transport sap. The heartwood is older sapwood in which the cells have become inactive. The heartwood often contains minerals that give it a darker color. In addition, the heartwood of ash, hickory, and some oaks have much of their porosity plugged by naturally occurring ingrowths. The mineral content and plugged pores can give heartwood uniquely different characteristics compared to sapwood. White oak heartwood is used for liquid storage barrels because of its natural sealing. Another attractive characteristic of heartwoods is their resistance to organic attack.

Wood is anisotropic and the strength in the direction of the grain is approximately 10 times higher than in the direction parallel to the grain. That is, the grains can be pulled apart much easier than they can be broken along their length. This characteristic is accounted for by good design practice or by layering sheared plies at right angles as is done with plywood. Wood also warps readily outwardly from the tree's core, as shown in Figure 4-1. Quarter-sawn lumber is the best cut because the grains run straight through the plank and will not warp.

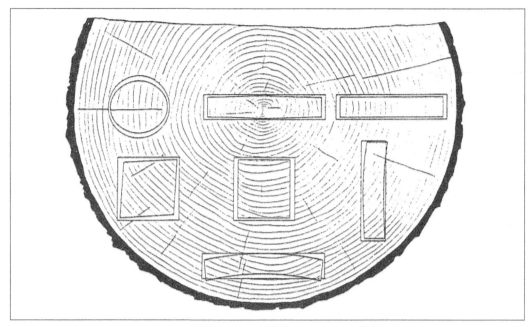

FIGURE 4-1. Wood warpage (image courtesy of USDA Forest Service, Forest Products Laboratory)

Determining the load-carrying ability of wood is based on many factors. Wood strength is decreased with the increasing presence of knots and other discontinuities (e.g., shakes, checks, and splits) as well as rot and moisture content. Again, the grain direction is the most important influence on strength. Laminated wood reduces the effect of knots by spacing them out across the wood and reducing the likelihood they will penetrate the entire laminate. The material properties of wood also vary based upon the type of wood and its moisture content. Oak is very strong in tension, and the tensile yield varies down through the hardwood and down into the softwoods such as pine and firs. White oak, for example, has twice the strength of Douglas fir in both shear and tension.

Wood is frequently glued to add tremendous strength to a screwed fastening. Glues or resins bond tenaciously to wood and distribute the joint loading over the entire joint rather than the area around the fasteners. Epoxy resins are excellent for bonding all types of wood, whereas polyester resin should not be used on redwood and close-grained woods such as oak or cedar.

Bamboo is a unique wood because its natural growth behavior produces a largely hollow structure. This makes it a natural tubing with a subsequently high strength-to-weight ratio.

Characteristics of Composite Materials

Composite materials are any substances made of two or more different materials. Natural materials are almost always composites, which are intricately mixed to make rocks, wood, soil, and nearly all other natural formations. Selecting complementary materials allows material scientists to customize composites for a particular application. Composites are typically understood to be a large-scale mixture of materials like the powerful combination of steel and concrete to make reinforced concrete. However, creative use of "composites" is also employed at the molecular level. Polymers and metals are often engineered with additives that add strength, ductility, and other desirable characteristics.

Composites used in fabrication generally consist of *fibers* and *resins*. The basic function of the fibers is to produce strength and stiffness while the function of the resin is to hold the fibers in place. In a more general form, fibers are referred to as reinforcements and resins as the matrix.

Composite materials, such as *fiber-reinforced plastics* (FRP), are stronger than the individual constituents because the resins are fluid and are able to transfer and distribute stresses among the fibers. FRP is a polymer-based matrix also known as a *polymer matrix composite* (PMC). There exist two other classes of composites based on the matrix material: the ceramic matrix composites (CMC) and metal matrix composites (MMC). Moldable plastics containing randomly oriented fiber reinforcements are also sometimes called "composites" although it is more common to refer to them as reinforced plastics.

Laminates, which are used to construct strong structural panels, consist of two or more layers of reinforcements, or plies, bonded by a resin. Sandwich construction is a special type of laminate construction that increases the rigidity and buckling resistance of the laminates by increasing the moment of inertia. By placing a thick but light and relatively weak core between two strong laminates, the assembly becomes much more rigid than if the two laminates were simply bonded to each other. The increase in rigidity is produced by the greatly increased moment of inertia resulting from the thickness.

Recall that for a given load, the deflection decreases with increased moment of inertia. Doubling the thickness increases the moment of inertia by eight ($I = bh^3/12$ for rectangular cross sections). The stresses in the core remain low under flexural loading because they are close to the neutral axis, therefore the c term in the flexural stress term is small ($\sigma = Mc/I$). Thus, making something thick makes it really strong and rigid and you can put cheap, weak material in the middle and it won't matter.

Commercial cores sold under various trade names include end-grain balsa wood, urethane foam, PVC foam, Kevlar honeycomb, and aluminum honeycomb.

CHARACTERISTICS OF FIBERS

Fiber material is the principal indicator of a composite's strength. This is why we refer to the composite by its fiber material. Fiber strengths are so high they are waiting for the resins and designs to catch up to their ability. The principal fiber materials for FRP construction are fiberglass, aramid (e.g., Kevlar), and carbon. These materials are sometimes mixed to provide a balance of strength and economy. Let's briefly review each of them:

Fiberglass

The fiberglass fiber varieties referred to as E-glass and S-glass are the most common structural fibers. E-glass is an alumino borosilicate glass and is the most commonly used glass in fiberglass-reinforced plastic. S-glass is about 33% stronger and 25% stiffer than E-glass but is also more expensive. Both fibers are white when dry and translucent when wetted with resin.

Aramid

Aramid (e.g., Kevlar) fiber was one of the first so-called "advanced" composites and is much stronger and stiffer than fiberglass. Aramid fibers derived from nylon are well known for their impact resistance. They are also excellent energy absorbers and fail in a ductile manner unlike the brittle, glass-like failures that occur in fiberglass and carbon.

Carbon

Carbon fiber is also considered an advanced composite, and due to its strong molecular bonding, is very solid and rigid. Stacks of hexagonal carbon plates are attached at the ends to form filaments. These tightly bound carbon plates make the filament strong. However, this same characteristic also produces a slippery surface and makes resin bonding difficult.

The mechanical properties of carbon are highly dependent on the nature of the parent material or *precursor*. As a result of this dependency on a specific precursor, there are a variety of carbon-fiber composites, each with differences in strength and stiffness. Current carbon-epoxy composites can achieve tensile yield strengths of 160,000 psi (1,100 MPa). Carbon has some other impressive features—for example, its strength increases with temperature and it has low thermal expansion.

One of the problems with carbon fiber is that its highly cathodic electrical nature can produce strong galvanic corrosion. Consequently, when metal fasteners are required, they must be made from stainless steels, inconels, or titanium.

Chacteristics of Adhesives

Most adhesives create a mechanical lock between surfaces. The adhesive works its way into the nooks and crannies of a surface and then hardens, creating a locking matrix. The adhesive is further connected by molecular surface forces (hydrogen bonding or Van der Waals' interaction) or shared electrons (covalent bonding).

Adhesives fall into one of these categories: wet, contact, or reaction. Wet adhesives, such as polyvinyl acetate (PVA or white glue), usually have a solvent that evaporates and creates intimate contact between two materials. These are best for porous materials. Contact adhesives are applied to all the materials being bonded. After an evaporation time, they can be united and will immediately bond. Reaction adhesives are usually two-part adhesives where one part is a catalyst that rapidly causes the adhesive to cure. Single-component versions may be triggered by such things as ultraviolet light.

Cyanoacrylate adhesives (CA), such as superglue, come in different thicknesses. They form a strong mechanical bond when the water present on the surface combines with groups of cyanoacrylates to rapidly harden. Van der Waal forces lock the bond. CA accelerants allow faster cure and the ability to adhere things with CA that might not normally adhere. CA accelerants are helpful when bonding wood structures such as balsa or basswood, but the smell can be especially bothersome. CA is brittle with low-impact resistance, but the instant connections are great for wooden model making.

The strongest bonds are made with polyurethane and epoxies. Polyurethanes produce mechanical bonds but also link to materials, such as the cellulose in wood, with powerful covalent bonds. Epoxies are two-part systems that become very strong and can be used to fill gaps. They can be mixed with fillers to make a putty or obtained in optically clear forms, which is helpful if you want to show off underlying materials. Epoxies are the strongest bonding agent for high-performance fibers such as carbon.

Polyvinyl acetate (PVA), such as craft glue, works in the same way but the connection to the materials is through weaker hydrogen bonds. Polyvinyl acetate is water based and relies on evaporation, and is therefore slow to cure. However, it is easy to clean up, has low toxicity, and is initially tacky so it can be more comfortable to work with. In addition, you can fill it with sawdust to make a cheap, sandable joint filler. PVA is also used as a fabric adhesive, which is great for prototyping soft goods, especially if you have poor sewing skills.

Hot glue is a simple and fast glue in which a polymer glue stick is heated and extruded onto a surface. It cools quickly and produces a mechanical bond between anything. This *hot melt adhesive* (HMA) technique is great for prototyping—it gives an ugly joint but you can count on it. However, the commonly used low temperature version is not very strong or temperature resistant. It is good to have a wet towel nearby to cool the inevitably burnt fingers.

Pressure -sensitive adhesives (PSA) and spray adhesives produce instant surface adhesion and are helpful for adhering graphics and surface veneers.

PVC cement is comprised of materials that dissolve PVC, especially tetrahydrofuran. Once the PVC is softened, cross-links are formed between the mating PVC materials. The solvents used to clean the PVC consist of tetrahydrofuran plus powerful cleaning solvents such as acetone and methyl ethyl ketones. PVC is a fast material for making large structural models, compressed air cannons, and even water lines.

Some specialty adhesives that are helpful for designers are fabric adhesives that can replace sewing and are especially helpful for building fabric models. Other specialty adhesives include thread locking adhesives, which are anaerobic and are designed specifically for fasteners.

Composite structures are made using adhesives, such as epoxy, urethane, or methacrylate. Methacrylate will work through oil films so less surface preparation is required. The layering of fiberglass and epoxy over a wood structure can give it new life, and is helpful for building high-strength, weather-resistant wonders. Figure 4-2 shows my new wooden boat covered with fiberglass and epoxy.

FIGURE 4-2. Wooden boat covered with fiberglass and epoxy

Corrosion Behavior of Materials

Corrosion ranks with fatigue as the most common causes of material failure. However, corrosion and fatigue usually work together: what corrosion starts is often finished by fatigue. Fortunately, the advent of plastics, stainless steels, and nickel alloys has provided great immunity from the continual corrosion process.

In 1780, the Italian anatomist Luigi Galvani hung some recently killed frogs from copper hooks. To his surprise, the frog's legs started kicking! As it happens, the copper hooks were connected to an iron rod. Fortunately, Galvani knew an electrical current would produce the same reflex action in the frog legs. Therefore, he connected the reflex action of frog legs with the unintentional and macabre phenomenon he observed on the copper hooks. This "animal" current, as it was originally called, led to experimental investigations into the remarkable phenomenon of *galvanic corrosion* and our current understanding of corrosion.

Because corrosion is a chemical process, it involves intricate discussion of electron behavior. However, there is a range of corrosion behavior from materials that never corrode, such as plastic and nickel alloys, to those that corrode with alacrity, such as steel. Between these extremes are metals that usually do not corrode but can do so under special circumstances, such as aluminum and stainless steels. The special circumstances are usually the wrong mix of metals, or exposure to acids and saltwater. Designers either need to avoid these situations or select the right material to accommodate them.

ANODES AND CATHODES

Corrosion results from two materials exchanging electrons. Metals have loosely attached electrons and will readily give them up to other materials. The material that gives up electrons, and consequently loses material, is called the *anode*, while the material that receives the electrons is called the *cathode*. The propensity of a material to give or receive electrons from another material is determined by the electrochemical nature of the material. The flow of electrons produces an electrical current and the use of anodes and cathodes is used to great advantage in batteries.

However, designers should avoid the corrosive reaction of metals by considering their galvanic relationship. This is important when the metals may become wet or are used in a humid environment. Metals that have similar electrochemical behavior generally will not cause each other to corrode. Some materials, such as aluminum and copper (including its alloys of brass) will cause corrosion if they are in intimate contact. Galvanic corrosion is eliminated if they are separated by a nonconducting material or if they are kept very dry.

RUST

Rust is caused by the iron in steel reacting with the oxygen. Oxygen is present in both air and water. The iron and oxygen molecules combine to form the nonwater soluble rust. The reaction sequence is described here:

1. $Fe \rightarrow Fe^{3+} + 3e^-$ (An iron atom dissolves, creating an iron ion and three free electrons. This reaction actually occurs in two steps.)

2. $4e^- + 2 H_2O + O_2 \rightarrow 4(OH)^-$ (Four free electrons coming from the iron reactions, two water molecules and one oxygen molecule react to form four hydroxyl ions.)

3. $Fe^{3+} + 3(OH)^- \rightarrow Fe(OH)_3$ (The iron ion reacts with three hydroxyl ions to form one insoluble rust molecule.)

Oxygen forces nearby metals (which readily shed electrons) to become anodic. Zinc and magnesium have a higher electrode potential (a relative measure of their naturally occurring voltage with respect to hydrogen), and therefore protect steel. Zinc and magnesium offer this protection because they readily donate electrons, with greater propensity than does the iron in steel. Because of their very loosely held electrons, zinc and magnesium are often used as sacrificial anodes in hot water heaters and boats. They corrode away instead of the steel tank or hull. These materials become unwitting accomplices in a designer's plan for corrosion protection.

The following zinc and magnesium reactions occur faster than the dissolving of iron that was previously discussed:

1. $Zn \rightarrow Zn^{2+} + 2e^-$ (A zinc atom dissolves, creating a zinc ion and two free electrons.)

2. $Mg \rightarrow Mg^{2+} + 2e^-$ (A magnesium atom dissolves, creating a magnesium ion and two free electrons.)

The electrons contributed by the zinc or magnesium then react with the hydroxyl ions produced by the solution of oxygen.

Some metals, including stainless steel, aluminum, and titanium, can be *passivated* by isolating the metal from the cathode. In stainless steels, the chromium is strongly attracted to oxygen and forms a protective oxide layer. In a similar manner, aluminum and titanium both react with oxygen to form a thin, protective oxide layer. This passivation reduces the rate of corrosion but does not eliminate it. The coatings are susceptible to chemicals that will strip the oxygen away and destroy the oxide coatings.

The following list includes common materials and their relative galvanic relationships in seawater. The further apart the metals are in the list, the greater the galvanic action and subsequent corrosion. For example, graphite is used with zinc in dry-cell batteries. The list shows some metals in their position when passivated. The loss of the passivation will make them much more anodic. This listing presents an average for alloy's galvanic relationship and subtleties exist among specific alloys, such as stainless steel:

Most anodic

- Magnesium alloys
- Zinc
- Aluminum alloys
- Carbon steel
- Cast iron
- Tin
- Brasses
- Copper
- Nickel-copper alloys
- Titanium
- Platinum
- Graphite

Most cathodic

This list illustrates some of the following points:
- Brass and steel hardware should not be mixed.
- Aluminum will slowly corrode around steel fasteners and quickly corrode around brass fasteners.
- Zinc and magnesium can be used to protect steel, brass, and almost any other material.
- Graphite is more cathodic than any other material it contacts.

Electrolyte conductivity accelerates galvanic action. Therefore, galvanic corrosion increases with increasing temperature, salinity, and pollution. The edges of bolt threads expose a lot of surface area and will quickly show galvanic corrosion.

CREVICE CORROSION AND PITTING

Small crevices and cracks between materials can restrict the free movement of an electrolyte and lead to an electrolyte that has a different composition in the far reaches of the crack. This system is called a concentration cell. Many different types of concentration cells exist depending on the composition of the materials. However, the oxidation-type concentration cell is one of the most common and is a typical cause of crevice corrosion.

Concentration cells are an important phenomenon because they produce corrosion in difficult to view areas such as underneath fasteners, clamped joints, scale, dirt, and cracks. Moreover, this accelerated corrosion in stagnant crevices can eat deeply into a small area of the material and make it weaker than the surface condition indicates. Crevice corrosion is a bigger concern with metals that rely on passivated (oxide) coatings for protection, such as stainless steel and aluminum.

Like crevice corrosion, pitting can deceptively weaken a material. Pitting is initiated without a crevice and occurs randomly over the surface of a metal. Pitting becomes more prominent than overall surface corrosion in metals that either have high alloy content or develop protective oxide coatings.

INTERGRANULAR CORROSION

Many alloys are also subject to intergranular corrosion. This corrosion shows as a web of cracks, typically made more obvious by the corrosion in the cracks. The composition of the grain boundaries may be different from the grain, resulting in a lack of protection for the material at the grain boundaries.

Microscopic galvanic corrosion can actually occur within an alloy where two different grain compositions are present such as in aluminum alloys. Corrosion can also be initiated by *stress cells* in which the stressed part of a material acts as the anode and the unstressed part acts as the cathode. This can be seen in metals formed by such processes as bending or stamping.

STRESS CORROSION CRACKING

Stress corrosion cracking is caused by a combination of tensile stress and intergranular corrosion. The material stress may be produced either externally or by residual stress produced during manufacturing (e.g., punching). Stress corrosion cracking creates the dangerously surprising condition in which a normally ductile material fails in a brittle manner.

BIOLOGICAL CORROSION

Biological corrosion or microbiologically influenced corrosion is a special form of corrosion resulting from interactions between a material and living organisms. Bacteria, fungi, and mold can excrete acids that chemically attack a material.

Designing to Prevent Corrosion

These are some favorable design features that reduce corrosion:

- Material selection should balance corrosion resistance and economy. Fortunately, plastics have allowed designs to be liberated from corrosion at a low cost.
- Avoid galvanic couples.
- When materials cannot be galvanically matched, fasteners should be more cathodic than their surrounding metal. Fasteners are much smaller than the material they connect. Therefore, if one of these items is going to galvanically corrode, it should be the massive fastened material because it can sustain the material loss better than the diminutive fastener.
- The anodic metal in a dissimilar metal connection should not be coated. A small break in the coating would allow the exposed anodic material to rapidly corrode under the influence of a much larger cathodic area.

CORROSION CHARACTERISTICS

The following list summarizes the most common forms of corrosion in various metals:

Steel

Unpainted or uncoated carbon steel will rapidly rust, especially when subjected to salt or other electrolytes. The rust is coarse textured and flaky. Polished and hardened steels corrode more slowly than unfinished steels. Many paints and coatings now permit steel to be used successfully in outdoor applications.

Stainless steel

Stainless steel is well known as being resistant to corrosion. While an oxide layer protects the metal, it is vulnerable to pitting and crevice corrosion. Stainless steels with a low chromium and low nickel content are the most susceptible to corrosion.

Cast iron

Cast iron will rapidly form a rust surface that bonds to the underlying iron. The bonded rust does not instantly flake off (as occurs with steel) and protects the iron from corrosion. The bonded rust is brown colored but can look almost black when submerged in water. Nickel and silicon cast irons have the best corrosion resistance, however all irons have very good resistance.

Aluminum

Aluminum is very corrosion resistant because of its strongly bonded oxide layer; however, it can be subject to both pitting and crevice corrosion. Aluminum is also very vulnerable to galvanic corrosion because it is very anodic. Therefore, it must not be in direct contact with more cathodic metals, especially copper alloys. This is especially a problem when an electrolyte is present. In dry environments galvanic corrosion between aluminum and other metals is uncommon. Corroding aluminum often produces a crusty, light-gray scale.

Copper alloys

Brass and bronze will readily corrode when in an electrolyte (like seawater) and in contact with steel or aluminum. Corrosion can appear in different forms from a rough green to a smooth brown appearance. Brass and bronze are also susceptible to the weakening effect of *dezincification*, which is the leaching of zinc from a material. Dezincification can produce a bright copper color. Brass will also react with ammonia, a common component in cleansers, producing a blue-green corrosion.

Nickel alloys

Nickel alloys are virtually impervious to corrosion. However, pitting is possible with some alloys.

Mind Story: Ski Boot

In Chapter 1, we briefly considered a helicopter blade as an example of hybrid materials construction. The blade has different material for different purposes, metal leading edges for surface strength, and honeycomb core for lightness around the neutral axis. These selections could not have been developed by a computer program. These selections required a textured understanding of the application and human wisdom.

Let's consider another application that goes beyond finding an optimal material: the ski boot. Ski boots have competing interests: rigid and strong soles, comfortable inside fit, adjustable clasps, rigid (but not too rigid) ankle support. Ski boots are an amalgam of plastics and metals. The outer shells are made from thermoplastic polyurethanes (TPU), polyolefin copolymers, and polyamide (nylon). The liner is made from ethylene vinyl acetate (EVA) foam. Rubber is used in the sole, and aluminum alloys for the buckles.

Closing Thoughts

You know materials are the raw ingredients for your design. Their capabilities inform how you design something. If you want something light and strong, you know that a carbon-fiber composite or aluminum are the best choices. You know they need to be processed differently and can be expensive. If it weren't for money, you'd probably design everything to be made out of titanium! You know if you need something to be cheap, you will consider the commodity plastics like polyethylene. Maybe you will consider cardboard or cast iron.

You know that materials are evolving, so "cheap" plastic has moved to high-temperature, high-strength, UV-resistant super material. You can even make them conductive with a range of textures, colors, and other appealing properties. However, you know there is an aesthetic appeal to wood, stone, and metals that is vital in some designs.

Materials can cause us grief. They can be beautiful to look at and touch, but they can be difficult to process or repair. They can have incredible strength, but be extraordinarily expensive. With so many material choices, you will probably end up with your favorites—a palette of materials that are almost perfect for every design.

CHAPTER 5

Thermodynamics

WHY DO WE STUDY THERMODYNAMICS? THIS MIGHT BE ESPECIALLY TOUGH FOR THE designer who loves to doodle, brainstorm, and create wonderful things. However, thermodynamics is the foundational discipline for looking at how energy moves. It provides a stepping stone for understanding how fluids flow and how heat moves. These are immensely practical concerns for those who want to design beyond the ubiquitous variants of an office chair.

In this chapter, we answer questions such as: Why does water boil? What makes ice skates work so well? And why does compressed air feel cool? Fundamentally, we are considering how energy moves from one form to another.

Thermodynamics is the science of energy and provides an explanation for the movement of energy. Thermodynamic properties can be observed from common phenomena such as the increased air pressure produced in a hot tire or the operation of a bicycle pump. Thermodynamic properties let us consider the nature of manifestations of energy, such as temperature, pressure, and velocity.

As hinted at previously, thermodynamics, fluid mechanics, and heat transfer are interrelated in many ways—they describe how energy and fluids move. For the conceptual approach taken here, it is best to digest all three of these topics at once and find the interrelationships. You will see the thermodynamics discipline provides a helpful way to look at energy, while fluid dynamics expands on how energy can change form in practical ways. You will see how fluid flow affects heat transfer dramatically. While there are also unrelated aspects of these disciplines, a completed circuit of energy and fluid concepts is beneficial in the world of design.

When considering thermodynamics, we conceptually place invisible boxes around systems and look at what is going in and what is going out. For example, thermal energy is driven across this system boundary by a temperature difference while work is energy driven across by other forces such as a pump or turbine.

The Baseline Temperature

Temperature measures the average vibration or kinetic energy of molecules and is given in units of Fahrenheit or Celsius. However, in thermodynamic and radiation heat transfer evaluations, we need a baseline for comparison. The baseline used is called absolute zero and is the temperature at which a molecule has no vibration energy at all. The absolute temperature scale in English units, where zero represents absolute zero, is:

R (Rankine) = °F - 459 °F

In SI units:

K (Kelvin) = °C - 273 °C

Note the degree symbol (°) is dropped for Kelvin and (increasingly) for Rankine units.

Pressure

We usually think of pressure as an external force over an area, so we know an ice skate with its sharp edge puts higher pressure on the ice than a tennis shoe. Same force, less area gives higher pressure. However, let's consider the internal pressure on a gas or liquid, which is comparable to stresses in solids. Static pressure is the force (per area) that keeps your tire inflated. Dynamic pressure, sometimes called velocity pressure, is actually kinetic energy, but it is convenient to think of it as the "pressure" produced by flowing fluids. The action of dynamic pressure can be seen most dramatically as the "water hammer" or surge pressure produced when a valve is closed quickly.

The SI unit for pressure is Pascal (Pa), which is equal to 1 N/m^2. Other common units are bar and atmosphere. In English units, pounds per square inch (lb/in^2 or more commonly psi) are used. The conversions of these units are 1 atm = 101,325 Pa = 14.7 psi or 1 bar = 100,000 Pa = 0.1 MPa.

Note psig represents gauge pressure—that is, pressure with respect to the atmosphere. Psia represents absolute or atmospheric pressure. The conversion between these units is:

psig + 14.7 = psia

Ideal Gas Law

The ideal gas law shows the relationship between pressure, volume, and temperature as such:

$PV = nRT$

where:

P = Pressure
V = Volume
n = Number of molecules
R = Universal Gas Constant
T = Temperature

This relationship only applies to an ideal gas—that is, a gas at a low pressure with no forces between molecules and where the molecules occupy negligible volume. The ideal gas law does not apply at high pressures such as would be encountered in an explosion or at high Mach number (described later). At high pressures, the molecules interact and their surface force attractions begin to alter the ideal behavior of a gas.

The ideal gas law quantifies commonly understood processes. The increase in pressure produced by heating a sealed cylinder can be described by the ideal gas law. Consider the behavior of an automotive tire as it heats up due to road friction. The added heat raises the temperature of the air in the tire and causes it to expand. But because the tire cannot expand very much, the volume is virtually unchanged. Therefore, the tire pressure increases as a direct function of the tire air temperature.

Another example of the ideal gas law is seen in the behavior of a bicycle pump. When the volume of air in the pump is decreased by pushing down on the cylinder, the air pressure is increased. The bicycle pump uses the increased pressure to force air into the bicycle tire. Consider also the simple process of blowing up a balloon: you are adding molecules (n) and the volume and pressure correspondingly increase.

Thermodynamic Laws

The laws of thermodynamics provide a framework for understanding the potential abilities of energy and mechanical work. These laws include some famous laws such as the Bernoulli principle, conservation of energy, and the second law of thermodynamics.

FIRST LAW OF THERMODYNAMICS

This first law states that energy is always conserved. This is an immensely important law that explains a wide variety of phenomena. The typical portrayal of the first law is a change in internal energy is equal to the addition of heat plus the addition of work (such as pumping energy). Internal energy of a substance is all the energy in all its forms. More expansively, the first law states energy can move between its many forms such as temperature, speed (kinetic energy), pressure (potential energy), work (mechanical energy, such as a pump), and heat transfer. A change in any one of these forms of energy will affect another form of energy in the opposite direction. Often systems are modeled as being *adiabatic*, which means there is no heat transfer into or out of the system.

Examples of Energy Change

While the first law refers to equivalence of energy, we should note energy can be added to a system. We can add heat with a torch, work with a pump, or kinetic energy with a catapult. If I have a closed loop system like a water filter for a fish tank, the whole system of the fish and filter are a closed loop. The pump is used to overcome the vertical height of the fill tube. That is, work equals potential energy. Now if I heat the water, put the tank and filter up on a top shelf or throw it across the room, I have changed the energy of the whole system. The whole system is affected because I am using an outside source of energy, not just changing the form of energy within the fish tank. The pump also has to overcome friction, both from the tubing and the filter. This frictional loss makes the system not absolutely adiabatic. The friction produces wasted energy that leaves the system.

An example of the first law of thermodynamics is when heat (Q) is added to a sealed cylinder and the potential energy (pressure) and temperature increase inside the cylinder. Consider a balloon filled with air—remember you increased the number of molecules (n) so the volume and pressure increased. When the air is released the balloon flies around madly, illustrating the conversion of potential energy (compressed air) to kinetic energy (balloon's speed).

Centrifugal pumps do the opposite—they accelerate a fluid with an impeller and then slow it down in the outer casing. This slowing of high-speed fluid creates high pressure. Another example, and one of the early experiments that proved the first law, is to add work to a system. Rotating a propeller inside a water tank will increase the temperature of the water in proportion to the work being done on the shaft. That is, the mechanical work is converted to temperature change.

Another example of energy change would be if cold air flows over a heating coil, heat is extracted from the heating coil, and the temperature (internal energy) of the water in the heating coil decreases. If one variable changes, all the variables change. For example, when heat is added to a car's cooling system (from the fuel combustion and friction) both the pressure and the internal energy increase.

Consider the example of air flowing through a car radiator, as shown in Figure 5-1. The incoming air absorbs heat from the hot water flowing through the radiator. This makes the air hotter and the water colder. The air is ejected into the atmosphere leaving the engine cooler.

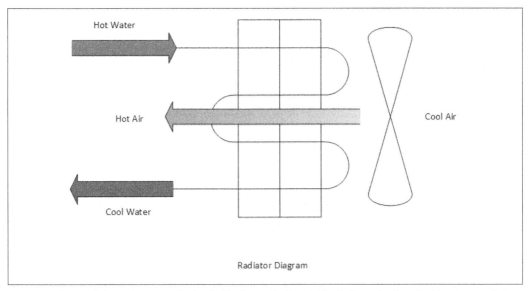

FIGURE 5-1. A car radiator exchanges water heated by the engine to outside air; the water flows inside a large surface area as a fan blows cool, outside air over these surfaces

A thermodynamic table is a convenient way to show how energy changes in this system. If we consider only the air, we observe that the temperature goes up and the heat transfer is into the air. The opposite is true for the cooling water. Although the air flows faster as it squeezes through the radiator fins, it slows down again when it exits. Therefore, the kinetic energy is unchanged. If we considered the state of the air as it was halfway into squeezing through the middle of the radiator, the kinetic energy would be higher than upstream or downstream values. This simplified diagram is a helpful way to see how the interrelationships work and it excluded frictional losses.

TABLE 5-1. Thermodynamic table for a radiator

Air (disregarding any fan work)	Q	T	KE	PE	W
Before radiator	-	-	-	-	-
After radiator	↑	↑	-	-	-
Water	Q	T	KE	PE	W
Before radiator	-	-	-	-	-
After Radiator	↓	↓	-	-	-

THERMODYNAMIC LAWS | 93

Abbreviations:
Q = heat
T = temperature
KE = kinetic energy
PE = potential energy
W = work

The complete first law equation can be written in several forms; however, if gravitation effects are ignored, one useful form that compares a pressure in velocity at any point 1 with any point 2 is:

$$P_1 - P_2 = \rho(V_2^2 - V_1^2)/2$$

where:

P_1 = Pressure at point 1
P_2 = Pressure at point 2
V_1 = Velocity at point 1
V_2 = Velocity at point 2
ρ = Fluid density

This is the Bernoulli equation, which quantifies the effect of velocity on pressure and explains airfoil behavior. When air is forced to travel a further distance over the curved top of a wing, the speed increases. According to the first law, when no heat is added or subtracted, another parameter must decrease to offset the increase in kinetic energy. In an airfoil, this increased speed drives down the pressure in accordance with the first law. Therefore, the pressure on the high-speed top surface of the airfoil has a lower pressure than the low-speed bottom surface of the airfoil. This difference in pressure is lift and this forces the airfoil up. The same principle produced the Venturi effect used in carburetors to draw fuel into an airstream.

Throwing a ball in the air is another example of converting speed (kinetic energy) to potential energy (attitude). Amazingly, most of the potential energy gets converted back to kinetic energy and returns the ball to you at the same speed you threw it. There is a slight decrease in speed because of the frictional losses to the air, which we describe in the second law of thermodynamics discussion and later in Chapter 6.

SECOND LAW OF THERMODYNAMICS

If energy can be moved between forms such as pressure, speed, and temperature, why do we need sun and food energy to maintain life or gasoline for our cars? Why does ice melt when placed in your hand? The usefulness or availability of energy as well as the direction of processes (heat moving from your hand into the ice) are described by the second law of thermodynamics. The second law gives a direction of energy flow, as shown in Figure 5-2. Energy flow direction is not contained in any fashion in the first law.

The second law states randomness (referred to as entropy) of a system always tends to increase. An example of this law is the observable fact that you can't throw the parts of a watch at a brick wall and have them land as an assembled watch. The energy recovered by a regenerative braking system that captures the braking energy won't equal the energy required to accelerate a car. This inequality is because friction causes heat energy to move into the roadway and the air.

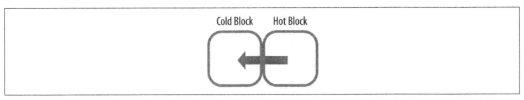

FIGURE 5-2. Energy flows in a manner that increases randomness; therefore, the hot block heats up the cold block

THIRD LAW OF THERMODYNAMICS

The third law of thermodynamics states internal kinetic energy or entropy of a perfect crystal at a temperature of absolute zero (−273°C) is zero. Absolute zero is the minimum energy condition of a material and it provides a basis for measuring and comparing entropy values.

ZEROTH LAW OF THERMODYNAMICS

To fill in a small gap in the thoroughness and logic of the thermodynamic laws, a zeroth law must be added. This law states that when two objects have the same temperature as another object (say a thermometer) the two objects are equal in temperature. This law is the basis for temperature measurements and is quite intuitive.

Nature of Phases

We think of molecules being in one phase (solid, liquid, or gas) but actually molecules are changing phases constantly and there is only a preponderance of molecules in one phase at any time. That is, liquids are evaporating at temperatures below their boiling point, as shown in Figure 5-3.

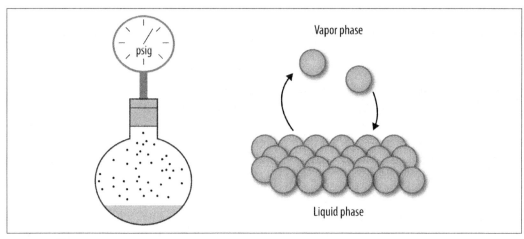

FIGURE 5-3. Vapor pressure

This phase changing is produced because the energy levels of the molecules vary. Internal energy or temperature is an average measurement. Some molecules have enough energy to escape the liquid phase and go into the gas phase. These evaporating molecules create a vapor pressure. When the pressure above the liquid is lower than this vapor pressure, the liquid will boil.

Another factor goes into considering evaporation as we ask the question: why does water go into a gas phase at room temperature? The answer lies in the fact that the higher-energy molecules leap into the gas phase but also that the air contains water and the water contains air. The relationship between solubility and pressure can be seen when opening a soda can. The pressure in the can lowers and suddenly forces the carbon dioxide to come out of the solution hissing and foaming in the process. Figure 5-4 provides an illustration of the internal energy of molecules and its relationship to a temperature measurement.

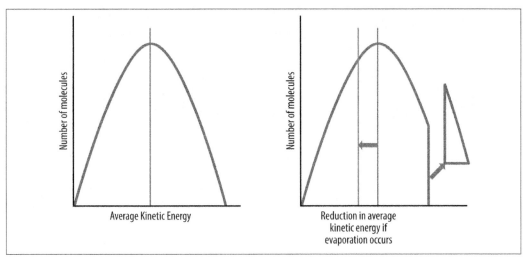

FIGURE 5-4. Nature of temperature

Notice the broad range of molecular energy; remember temperature is an average value. This curve also illustrates the effect of evaporative cooling. As the high-energy molecules to the right evaporate, they lower the average temperature of the remaining liquid phase molecules. This may beg the question of why high-energy molecules are in a liquid state, and the answer can be considered using a magnet analogy. Consider two magnets that are attached. You can separate the magnets by pulling them apart with a force greater than their attraction. This is what happens when molecules vaporize. However, there is a small time when even though the force on the magnets is greater than required, they have not separated. This is the condition of these high-energy molecules—give them time and they will vaporize.

In terms of real molecular surface forces, the attraction between the molecules is developed by the electrical attraction produced by the shape of the molecule. That is, most molecules are not spherical—rather they are all sorts of weird shapes dictated by the element's orbital shape. Water, for example, is shaped like a "V" with one end having a positive charge and the other end a negative charge. This polarity encourages water molecules to stick together (and makes them easily rotated by microwave ovens!).

Figure 5-5 provides insight into the nature of these three phases and how it relates to surface force adhesion.

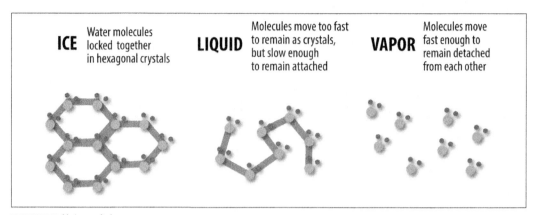

FIGURE 5-5. Nature of phases

A couple of elements, namely hydrogen and helium, are spherically shaped and non-polarized. Therefore, they must be nearly devoid of all thermal energy in order to stick together in a liquid or solid phase. They really act like ball bearings and don't want to stick to anything—hydrogen freezes at 14 K and helium at less than 1 K!

Because molecules are constantly transferring back and forth between liquid and solid phases, adding a foreign material, such as salt, interferes with the refreezing of water and lowers the freezing point.

Consider Figure 5-6 and what it says about the effect of pressure on phase change. The chart is called a pressure-enthalpy chart where enthalpy is the specific name for energy. You will note the liquid vapor lines go to the right (higher enthalpy) with increasing pressure. This is because the liquid phase molecules require higher energy to "jump" into the vapor phase. It is like a heavier weight pushing down on the molecules preventing them from taking this "leap of phase." Of course the same is true with the solid-liquid interface. The phase lines connect at very high pressures; this point is known as the triple point where all three phases coexist. We normally don't operate any equipment at these pressures, but it does close the loop on the pressure-enthalpy charts.

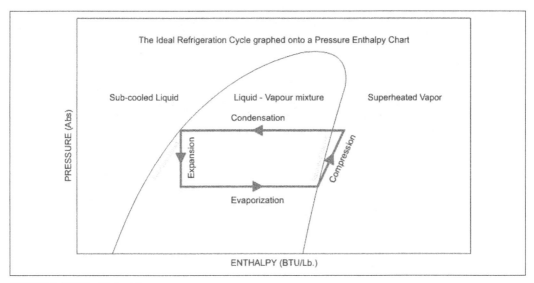

FIGURE 5-6. Refrigeration cycle

An interesting note in terms of phase change is that higher pressure forces a material into its higher density phase. In the case of water, this happens to be liquid rather than solid. That is, high pressure will force ice to become liquid, hence ice skates melt ice.

When water molecules get colder and therefore slow down, they occupy less volume. That is, water gets denser as it cools, which is common for most materials. However, as the water turns to ice, it starts getting less dense. This lower density of ice compared to liquid water is the reason that ice floats, as shown in Figure 5-7. Water is strange!

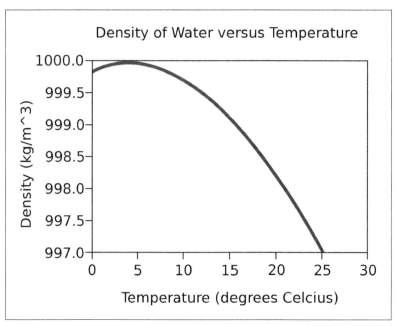

FIGURE 5-7. Water density versus temperature

This is confusing stuff, but has immensely practical consequences. This description of phases should harmonize an understanding of phases and how they behave. But it is still confusing to think of a liquid vaporizing at room temperature and 1 atm. It is also difficult to understand why salt depresses the freezing point of water.

Thermodynamic Explanation of the Cooling Produced by Expansion and Evaporation

Perhaps you have felt the discharge of a compressed air hose and noticed the air was cooler than the surrounding air. Cooling is produced when a fluid (gas or liquid) is expanded. Interestingly, some gases, such as hydrogen, helium, and neon, will heat up when expanded. Moreover, ideal gases (remember those theoretically noninteracting spheres?) will have no temperature change when they are expanded. So why do some gases cool and others heat up when expanded? The straightforward answer is that gases with a positive Joule–Thomson coefficient will cool when expanded.

The Joule–Thomson coefficient describes the relationship between a change in pressure and a change in temperature, as shown in Figure 5-8. Those materials with a negative value will heat up and those with a value of 1 will not change. The coefficient will change based on temperature and pressure. Therefore, all gases will exhibit a negative value at a high enough temperature and pressure. This cooling phenomenon at low temperatures and pressures has many practical applications, especially for liquefying gases.

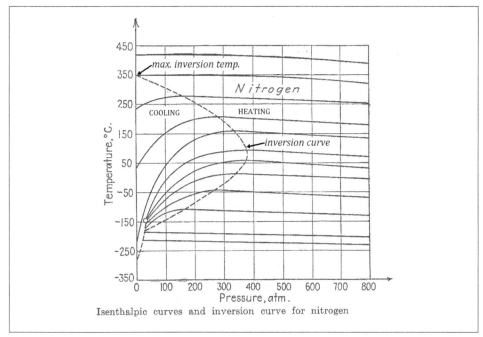

FIGURE 5-8. The curve slopes are the Joule–Thomson coefficient

However, in order to understand the cause of this change in Joule–Thomson coefficients as they relate to the cooling (and heating) phenomenon, we must look to the first law of thermodynamics and the van der Waals equation, which describes intermolecular forces.

First, consider the van der Waals equation. This equation takes the ideal gas law and includes the repulsive and attractive molecular interactions and the nonzero volume taken up by the molecules themselves. The equation is

$$p = \frac{nRT}{V - nb} - a\left(\frac{n}{V}\right)^2$$

where:

p = pressure
V = volume
n = number of moles
R = gas constant
T = temperature

NATURE OF PHASES | 101

The repulsive interactions and nonzero size of the molecules are taken into account by supposing the molecules are only free to move in a volume of $V-nb$ where nb is approximately the volume occupied by the molecules themselves. The pressure exerted by the gas on the walls of the container with a volume of V is related to the number and frequency of the molecular collisions with the wall. These are both reduced by attractive forces between the molecules and this is included in the $-a(n/V)^2$ term. The van der Waals equation requires two experimentally obtained coefficients, a and b. If a and b are known for a gas, then the equation can be used just like the ideal gas law to calculate p, V, n, or T.

Tables of values for a and b are found in reference books. Following are some values for a and b. The units for a are $liters_2$ atm. $mole_{-2}$ and the units of b are $liters$ $mole_{-1}$.

Molecule	a	b
H_2	0.2444	0.02661
O_2	1.360	0.03183
N_2	1.390	0.03913
CO_2	3.592	0.04267
Cl_2	6.493	0.05622
A	1.345	0.03219
Ne	0.2107	0.01709
He	0.03412	0.02370

This cooling associated with most expanding gases is caused by the increase of potential energy in the system produced by the increased distances between the molecules. The further molecules are separated, the higher the potential energy associated with their intermolecular attractive force. These forces are associated with van der Waals attraction as well as hydrogen bonding. Some molecules behave like magnets that have been separated—the further apart they are separated, the greater the total potential energy. Even though the force is weak at long distances, the total potential energy can be thought of as the summation of all the forces applied over the whole distance that separates them.

During expansion, the potential energy increases so the internal temperature must decrease in an adiabatic system with no work (that is: $\Delta Q = 0$, no heat added or taken away and there are no fans or turbines) to abide by the first law of thermodynamics. The change in temperature is a function of the intermolecular forces (reference the van der Waals equation) and varies by material. The temperature effect, which is typically a cooling effect, is quantified by the Joule–Thomson coefficient.

> Note that gases have a higher potential energy than do liquids because of these intermolecular forces. When gases condense, they release latent energy (as do liquids when they freeze). In fact, condensation occurs when the temperature of a gas is decreased so the molecules don't vibrate as much and get closer to one another.

Combustion

We all seem to have a primordial attraction to fire and a childlike attraction to explosions. Combustion offers good insight into how energy moves and the notion of chemical energy being another part of our energy consideration. The combustion of hydrocarbons requires mixing fuel with air and providing an ignition source. In chemical terms, hydrogen reacts with oxygen to produce water, while carbon reacts with oxygen to produce carbon dioxide. The combustion of the simplest hydrocarbon, methane, produces carbon dioxide, water, and energy. This is typically shown as:

$$CH_4 + O_2 \rightarrow CO_2 + H_2O + \text{Energy}$$

This reaction occurs in a sequence of events that produce unstable molecules until the reaction is complete. Therefore, complete combustion requires some finite time to occur because the sequence of intermediate reactions needs time to occur. In addition, the fuel and air have to be mixed thoroughly so all the fuel is burnt off.

However, incomplete combustion of a hydrocarbon (shown as C_xH_y) will provide unburned fuel, for example:

$$C_xH_y + O_2 \rightarrow CO_2 + H_2O + C_xH_y \text{ (unburned fuel)} + CO \text{ (carbon monoxide)} + \text{Energy}$$

This reaction shows the combustion of a generic hydrocarbon and pure oxygen to produce carbon dioxide, water, and unburned fuel or carbon monoxide.

Finally, combustion requires activation energy (or ignition source). That is, the fuel must be elevated to a certain temperature before the reaction will occur. The heat generated by the reaction will sustain further reactions. For example, a match is required to light a candle after which the combustion is self-sustaining. Diesel engines use the heat of compression for their ignition source, whereas petrol engines employ a spark plug to provide the ignition temperature.

When the ratio of fuel and air is such that all the fuel and all the oxygen are consumed, the mixture is referred to as stoichiometric. A stoichiometric reaction gives the highest efficiency and flame temperature because there are no unconsumed molecules to absorb heat. However, if the air supply is slightly reduced, dangerous carbon monoxide and unburned fuel can result. Consequently, engines are normally set to burn a little excess air to ensure complete and safe combustion.

You may wonder why some things explode and some only burn. The difference between detonation and deflagration is the rate of advance of the flame boundary. When the fuel and air mixture is ignited, the flame will propagate by either deflagration or detonation. Deflagration produces a subsonic advance of a flame front while detonation produces a high-pressure supersonic wave with a velocity up to 6560 ft/s (2000 m/s) and a pressure up to 290 psi (2 MPa).

Exceeding Mach 1, the speed of sound, causes the gases to no longer act like a gas but rather as a liquid, causing great destruction. Explosives produce a rapidly advancing wall of highly compressed air around the exploding fuel. Where ignition occurs in an open area, the flame front promotes turbulence in the unburned fuel/air ahead of it and therefore causes it to ignite more rapidly. This increased burning rate increases turbulence and subsequent flame velocity. The reinforcement of combustion can turn deflagration into detonation.

Certain fuel-to-air ratios will produce explosion. These values are well known so combustion system can either encourage or discourage explosions. For example, hydrogen gas will explode if the fuel-to-air ratio is anywhere from 4%–75%, acetylene is 2.5%–81%—scary stuff as those of us who have used similarly explosive oxy-acetylene torches have learned.

Some explosions are especially violent because they not only release a lot of chemical energy but the combustion produces many more molecules. For example, the explosion of nitroglycerin takes one giant molecule and creates seven new ones. Each one has high thermal energy and produces pressure. While the ideal gas law does not apply directly to high Mach flow, the relationships still exist. This relationship indicates that inserting more molecules drives up the pressure.

Because air consists of approximately 21% oxygen and 79% nitrogen, a great deal of nitrogen is included in the combustion zone. Normally, this lowers the flame temperature as the inert nitrogen absorbs heat but does not contribute to the reaction. The cooling effect of nitrogen also explains why pure oxygen produces a hotter flame for welding. At very high flame temperatures, however, nitrogen will react with oxygen and produce nitrous oxides (NOx), as shown by the following example of methane combustion:

$$CH_4 + (O_2 + N_2)air \rightarrow CO_2 + H_2O + N_2 + \text{a little bit of NOx}$$

Why do we care about nitrous oxides? The nitrous oxides react with oxygen and water to produce corrosive nitric acid. Consequently, the nitrous oxide production goes up steeply with flame temperature, but it can be greatly reduced by lowering the flame temperature or using a catalyst, like the ones used in catalytic converters in automobiles. Flame temperature is usually lowered by recirculating exhaust gases back into the flame. Also, nitrous oxides can be produced by nitrogen in fuel.

Sulfur dioxide and trioxide (SO_2 and SO_3, respectively) are other common combustion pollutants. These two pollutants, often referred to as SOx, are produced by the reaction of sulfur in fuel with oxygen. Like nitrous oxides, sulfur oxides will react with oxygen and water to produce an acid—sulfuric acid, in this case.

Also, unburned fuel and carbon monoxide can be produced due to poor combustion caused by insufficient time, turbulence, or substoichiometric conditions (too low fuel-to-air ratio).

Other pollutants are also generated by the combustion process. These include small quantities of toxic dioxins, furans, and mercury, as well as particles composed of unburnable inorganic materials such as silicon. The category of air pollutants that can cause serious health effects such as cancer and birth defects are called air toxics.

FUELS

Some common fuels are fuel oil, diesel fuel, gasoline, propane, alcohol, and kerosene. Light fuels such as gasoline, propane, alcohol, and kerosene burn easily. Heavy distillate fuel oils and medium distillate gas oils are safe in that they do not burn too easily and remain liquid over a broad range of temperatures.

Data sheets are available for all fuel, lubricants, and many other materials that provide a description of required safety equipment, clean-up procedures, flammability characteristics, and other safety concerns.

Terms used to describe fuel properties include:

Viscosity
Describes the relationship between flow and force. It describes how rapidly a fluid flows. Viscosity is a function of temperature and sometimes of pumping rate.

Flash point
The temperature at which combustion is supported by vapors.

Cloud point
Temperature at which waxes start to form and oil looks cloudy.

Gel point
Temperature at which fuel begins to turn into a jelly mass.

Octane
Describes the amount of compression allowable before ignition.

Cetane
Describes the ease of ignition for diesel fuel (without ignition aids).

Mind Story: Evaporation

Understanding why a cup of water evaporates takes a deep understanding of molecular behavior. First, you have to understand that all the molecules are not at the same temperature. Measured temperature is the average energy level. Many of the molecules have a much higher energy level (or temperature) than the average value. These thermal vibrations are so high that they break the weak surface bonds that forced the molecules into the liquid phase. Once these surface bonds are broken, the water becomes a gas and it wafts off into the room, taking its energy with it.

You can think of evaporation in terms of thermodynamic laws also. Because entropy (randomness) always increases, solids tend to turn to liquids, and liquids into gases. The increase of randomness is offset by a decrease in heat content (enthalpy). In addition to phase change, entropy is also increased by such things as increasing temperature, decreasing pressure, and moving from an undissolved form to a dissolved form.

Usually there is an equilibrium principle whereby for every group of high-energy molecules that jump into the gas phase, an equal group of low-energy gas molecules stick to each other and become a liquid. This condensate keeps the liquid level at the same point. This phenomenon occurs when you put a lid on a cup. When there is no lid, the molecules that turn to gases move away from the cup. The cup of water has been cooled down slightly by the loss of this molecule but it absorbs energy from the room to get back at room temperature. The higher energy molecules continually move into the gas phase and are not replaced with condensing water coming from the air. Eventually, the cup is emptied.

Mind Story: Dew Point

Everybody likes to talk about the weather, but this time let's do something with it. Many of us have enjoyed the glistening dew drops on summer grass or contended with its evil cousin, frost. Once the nighttime air has lowered to the point that water starts condensing, the condensation puts the brakes on temperature drop. Therefore, the lowest temperature in the evening is the dew point temperature. Once the temperature drops to the dew point, latent heat must be released to the air for the condensation process to take effect. This addition of heat offsets some or all of further cooling.

If the dew point happens to be much lower than the air temperature, the air will cool off much more rapidly at night than if the dew point was closer to the air temperature in the evening. This is why dry regions such as deserts have such large differences between the high and low temperature. Moist areas, such as regions near the coast, tend to have a smaller spread between the high and low temperature.

If dew is very heavy in the morning, the temperature will not rise as quickly when compared to situations when there is no dew. This is because as dew evaporates, it cools the surrounding air (absorbs latent heat). Eventually the sunshine will overpower the cooling produced by evaporation, but the temperature will be cooler than a location that did not experience any dew.

We are not done with weather yet. We will look at it as an example of radiant heating in Chapter 7.

Closing Thoughts

Thermodynamics is the study of moving energy and this is often the fundamental challenge in both mechanical and electrical design. You know that moving energy comes at a price: the transfer efficiency causes us to waste energy, often in the form of mechanical friction and electrical resistance. In mechanical systems, you know there is a common exchange between potential and kinetic energy. Pressure and velocity dance with each other, producing wonderful maneuvers of flying wings and impact wrenches.

We often tap into chemical energy from gasoline in our cars to natural gas and coal for power plants. Fuel is very energy dense, so it acts like a chemical battery to power all sorts of things. You know combustion can cause problems besides the obvious heat and explosion issues. Combustion usually produces some sort of pollutant, the most dangerous of which we call air toxics. While fire has a visceral appeal, there is a lot of chemistry going on inside that flame!

CHAPTER 6

Fluid Mechanics

WHILE WE STAND ON SOLID GROUND, WE LARGELY LIVE IN A WORLD OF FLUIDS (LIQUIDS AND gases). We walk around in a mass of air. We breathe it in our lungs and feel its pressure on our bodies. Blood flows through our body, keeping us nourished and healthy. We drink lots of water and occasionally swim in it. In the mechanical world, we use fluids like hydraulic oil to move things or we force things like cars to move through the atmosphere. We might wonder why boats and birds are streamlined, why golf balls are dimpled, what drag is, and how it's possible to stop a car with a touch of the brake pedal. Designers frequently work with wide-ranging fluids, including water, air, oil, and blood. Their difference in behavior can be surprising.

The study of fluid behavior is broken down into two disciplines. The first is the study of fluids at rest, called fluid statics. And second is fluid dynamics, which describes the nature of moving fluids.

Fluid Behavior

Solids differ from fluids in that a fluid cannot permanently resist shear stresses. Therefore, where a solid will deform (strain) a limited amount when a force is applied, a fluid will not. The rate of movement is dictated by the force but not the total amount of movement—it just keeps on flowing. A fluid eliminates any shear stresses simply by flowing.

The resistance to flow (technically, shearing force) is called viscosity. High viscosity indicates the fluid moves slowly and takes a long time to remove shear stresses and therefore acts more like a solid. Many fluids are so viscous that they need to be heated in order to pump or atomize (molasses is one such example). Fluids are also distinctive in that pressure is transmitted equally in all directions and normal to any plane. However, pressure disturbances can be imposed upon a fluid. These disturbances move as waves that travel at an acoustic velocity, which is a function of two material properties: density and bulk modulus elasticity. Bulk modulus elasticity represents a material's resistance to compressibility.

The inability of a fluid to handle shear stresses is actually the best definition of a fluid. Some apparently solid materials have fluid properties; that is, they gradually flow when a pressure is applied to them. In perfectly elastic materials, such as a rubber band or spring, the deformation occurs instantaneously. However, if a material is viscoelastic, it provides two different behaviors. It will elongate instantly but will then slowly flow like a viscous material.

You can measure viscoelastic behavior by quickly pulling on a material and measure the force to hold it in this stretched position. For viscoelastic materials, this value will decrease over time as the material rearranges itself. These viscoelastic materials are great at absorbing shocks and are used for such things as helmet padding. Not surprising. The cartilage that holds your bones together is viscoelastic.

A physical model of viscoelastic behavior can be developed using a combination of dashpots and springs, as shown in Figure 6-1. A dashpot is a damper that resists in proportion to velocity. It can be thought of as a piston inside a cylinder with some contact resistance so that as the piston sliding in the cylinder always drags.

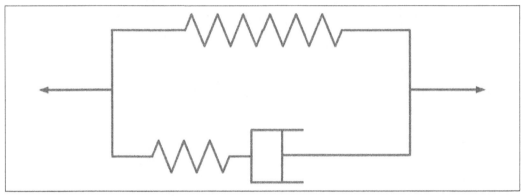

FIGURE 6-1. Models of viscoelastic behavior

Fluid Statics

The first law of fluid statics states that a fluid contained in a vessel, like coffee in a mug, will distribute an external pressure uniformly throughout the fluid. That is, a fluid will not have some portions in compression and others in tensions as occurs in solids. This law, referred to as Pascal's law, applies only when the pressure produced by the fluid's weight is neglected. In the case of hydraulic systems, this law would indicate that when one point increases in pressure, all points increase with the same pressure.

Water lies flat because the force of gravity is applied uniformly over its surface. If water were lifted up above the surrounding water, the weight of the elevated water cannot be supported by the surrounding water (specifically, the water cannot carry any of the shear stresses). Therefore, the water drops down to the same level as the remaining water. The resulting surface, called the free surface, can be affected by other external forces, besides earth's unrelenting gravity. Wind is the principal culprit in making waves, and the gravitational pull from the sun and moon creates the tides. Additionally, water current, seismic energy, geothermal energy, and even surface tension all affect the free surface of the oceans and other bodies of water.

The pressure experienced by a water molecule increases with depth. The molecules below the free surface not only experience the earth's gravitational pull (which is reduced by less than 1% at the bottom of the deepest oceans) but all the weight of the water above it. The gravitational pull on all the air above the free surface is small, only 14.7 psi (101 kPa or 1 bar), but the gravitational pull of the denser water molecules is much higher. The density of fluids also changes with their elevation, but it is tiny in liquids and is usually ignored.

Pneumatics and Hydraulics

Pneumatic and hydraulic system performances are dictated by pressure and the surface area upon which the pressure acts. Hydraulic and pneumatic systems rely on fluid static principles as embodied by Pascal's law, which dictates that pressure is equally distributed throughout a container. When pressure increases at one end of a hydraulic circuit, the pressure will instantly transmit to the remote ends of the circuit. Moreover, the force exerted by the fluid can be multiplied by taking advantage of surface area.

The relationship between area and pressure is illustrated in Figure 6-2.

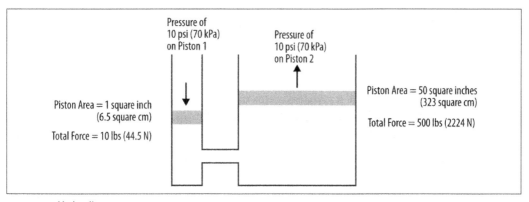

FIGURE 6-2. Hydraulic power

The relationship between area and pressure can be illustrated by considering the two hydraulically connected cylinders shown in Figure 6-2. We can see that because the pressures are identical, the force produced by Piston 2 is 50 times that produced by Piston 1. Because

$P = F/A$

where:

P = Pressure
F = Force
A = Area

And, for all the pressures being equal in the system in accordance with Pascal's law, then:

$$\frac{F_1}{A_1} = \frac{F_2}{A_2}$$

and:

$$F_2 = \frac{F_1 A_2}{A_1}$$

It is important to note the speed of operation is not part of this relationship. Speed is a function of fluid flow. For example, increasing the relief pressure of a hoist will not make it operate faster, it will only increase the power or work the hoist can generate.

Practical aspects of these pneumatic and hydraulic systems will be described in Chapter 10.

Fluid Dynamics

We live off the fluids flowing through our bodies; we walk through and breathe a fluid. Because we interact with flowing fluids in such a routine fashion, they are a wonderful topic of study that we call fluid dynamics. We could consider the entire fluid mechanics discipline as fluid dynamics with fluid statics being a special case where velocity is equal to zero. We could even think of fluids as being like solids, but with a really low shear strength. However, when fluids flow, all sorts of amazing things happen as pressures and velocities develop in peculiar ways.

RHEOLOGY

Fluids can respond to forces in surprising ways. We see this through rheology, which is the study of how fluids move under an applied stress, such as being pumped. We are familiar with water's behavior—the faster you push your hand through the water the more resistance you feel. The behavior of speed and resistance has a nice linear relationship that is produced in what are called Newtonian fluids. Most fluids are Newtonian with their direct relationship between viscosity and flow rate. The shear rate can be practically thought of as the pumping speed.

However, some fluids are non-Newtonian and behave differently at different speeds, such as molasses. The common categories for non-Newtonian fluids are pseudoplastic, dilatant, and Bingham plastic. Because their viscosity is dependent on their shear (or pumping) rate, pseudoplastic fluids are also called shear-thinning and dilatants are called shear-thickening. These fluids are common in biological systems and have practical applications for power transmission, traction control, and body armor.

Liquids that are dilatant will flow easily at low speeds, but become very viscous at high speeds. A mixture of corn starch and water is the classic example of dilatant behavior. You can drag your finger slowly though and it feels like water. However, it becomes viscous if you punch it. Silly Putty is another example. You can easily stretch it, but if a load is quickly applied (a high shear rate) such as dropping it on the floor, it becomes highly viscous (like a solid) and bounces.

Dilatant behavior is typical when high concentrations of solid particles are floating in a liquid. This combination of solids and liquids is called a slurry. As slurry flows at low speeds, the whole assembly of solids and liquids flows easily, but at higher speeds the particles can't move fast enough and makes the whole fluid assembly highly viscous.

Pseudoplastic liquids are the opposite of dilatant. They will flow easily at high speeds but become very viscous at low speeds. This is great for food! It is helpful that whipped cream can easily flow at high speeds, so too that cake mix flows nicely around spinning beaters. Molasses is another example of this behavior; it is really hard to start it moving but once you do, it is easy to flow.

Toothpaste and blood are examples of a Bingham plastic. Toothpaste won't come out of the tube until it is squeezed. However, once the toothpaste is flowing it acts like a Newtonian fluid.

Blood is comprised of cells suspended in water-based plasma. Blood has complex flow in the body because red cells are deformed by collisions and their ability to deform allows them entry in certain capillaries. Moreover, blood will stop and start in some veins while it will flow rapidly in arterial flow. Blood at very small forces won't move at all, but once a certain force is exceeded it will flow. This flow behavior is due to the solid blood cells being packed into the blood and it takes a force to get them moving. Even when not coagulating, the red cells tend to form structures that resist flow.

> ## Mind Story: Blood
>
> Blood is comprised of a liquid and solids of different sizes. It is a non-Newtonian fluid that behaves differently at different speeds and pressures. The liquid plasma is 90% water, and the solids are comprised of platelets, red cells, and white cells. White cells are the biggest solid and platelets are the smallest.
>
> When blood flows quickly through a wound, the liquid plasma flows rapidly while the solid cells lag behind and start bunching up. The surface forces of the cells make them stick. The disk-shaped platelets initiate the clot. Proteins in the plasma create a mesh of fibrin to strengthen the platelet clot. This amalgam stops the bleeding.
>
>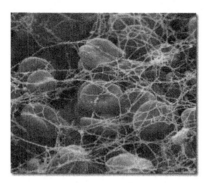

BOUNDARY LAYER

When a fluid flows over a solid, the fluid is slowed down. At the surface of the solid the fluid sticks and has no velocity. Further away from the surface, the fluid gradually flows to its full velocity. The transitional area between where the fluid sticks to a moving object and where the fluid has developed its speed is called the boundary layer.

Two types of boundary layers can develop: laminar and turbulent, as shown in Figure 6-3. In water, only small, slow-moving objects have laminar boundary layers. Turbulence is a mysterious contortion of a fluid that flails about in unpredictable ways. However, contrary to the pejorative notion of turbulence, when considering the tiny area around a surface where the boundary layer exists, a turbulent boundary layer is advantageous. A turbulent boundary layer is thinner and binds better to a flowing object than a laminar boundary layer. This stickiness of turbulent boundary layers reduces drag.

Boundary layer thickness is related to both drag and heat transfer (which will be discussed in Chapter 7).

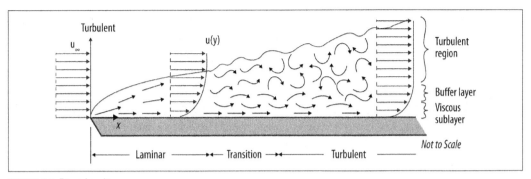

FIGURE 6-3. Boundary layers

Dimensionless Parameters: Mach and Reynolds Numbers

The behavior of flowing fluids is related to dimensionless parameters that relate the inertia of a fluid to some other parameter, such as viscosity. A couple of helpful dimensionless parameters are Mach and Reynolds numbers.

The Mach (Ma) number indicates if a gas has compressible or incompressible flow. High Mach flows are compressible and are handled differently than low Mach, incompressible flows. Mach number is the ratio of actual velocity to the speed of sound through the fluid:

Ma = V/a

where:

V = velocity of the gas
a = speed of sound through fluid

The speed of sound is the velocity of mechanical compression and rarefaction through a fluid; the speed of sound gives Ma= 1, which is 740 mph (330 m/s) in air. The density of low-velocity gases does not vary a great deal while for those approaching Mach 1 (the speed of sound), compression pressures cannot be distributed throughout the gas. The mathematical treatment of fluids is divided between those that can be compressed (gases traveling at low Mach) and those that cannot (high Mach gases and liquids).

The Reynolds number (Re) is a measure of the ratio of inertia to viscosity. Thinking of velocity instead of inertia in the numerator may be more intuitive. The equation takes many forms but fundamentally Reynolds number is calculated as:

Re = Inertia/Viscosity

Therefore, high-velocity flows will have a high Reynolds number while low velocity will have a low Re. Because the Reynolds number is inversely related to viscosity, high-viscosity fluids (such as oil) will have a lower Re than gasoline even if both are flowing at the same velocity.

The Reynolds number is used to determine drag and identify whether the flow is laminar or turbulent. Regardless of the fluid, Reynolds numbers less than 2000 typically result in laminar flow, while Reynolds numbers over 2300 produce turbulent flow. A transitional range exists where the flow can be either laminar or turbulent. You can see the transition from laminar to turbulent flow by running a stream from a faucet at a low flow rate and then increasing the flow until it trips into turbulent flow.

DRAG

For a designer, understanding drag is probably the most important practical consideration in fluid dynamics. Total drag is the summation of three types of drag: friction, pressure, and wave making.

Water dragging against a pipe is an example of friction drag and this can be reduced by smoothing the surface or reducing fluid velocity. Pressure drag is produced by geometric changes because fluids don't like changing direction. Therefore, pressure drag is produced when fluids are forced to do so. Consequently, gradual turns produce less pressure drag than sharp turns. You can feel the effect of pressure drag by putting your hand outside a car window while driving and noting how the resistance varies with your hand shape: open facing the wind, open in line with the wind, and a fist, for example. The same results can be seen by putting your hand under a gushing water fountain—pressure drag is a giant force and usually is the major drag component in many systems.

One example of pressure drag is seen in boats, as the water is forced to part around the hull. As shown in Figure 6-4, hull pressure drag results from the pressure difference between the front of the hull and the stern where wake eddies exist. Streamlining the hull is vitally important in reducing pressure drag. A hull that is in the shape of a box (flat bow and stern) will have 15 times as much drag as a streamlined hull (rounded bow and stern) of the same width. With a squared off bow and stern, the bow impacts the water creating high pressure, and then radically disengages the water at the stern, thereby producing powerful eddies. These eddies result in very low pressure behind the hull.

The combination of flat bow and stern features creates a large pressure differential and results in pressure drag across the hull. The streamlined hull gently changes the water direction around the bow and decreases the pressure at the bow. More importantly, it allows the water to recombine astern with a minimum of eddies and subsequent low pressure. Therefore, the streamlined hull produces a lower pressure at the bow and a higher pressure at the stern. Consequently, the pressure differential across the hull is much lower, creating a lower pressure drag.

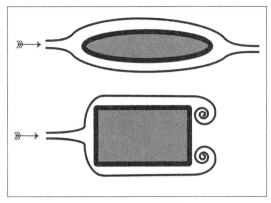

FIGURE 6-4. Hull drag forces

Drag Coefficient

The drag coefficient (Cd) is a convenient nondimensional parameter for relating drag to shape. Figure 6-5 provides a listing of drag coefficients for different shapes. The calculation is described by the following equation, which shows the linearly proportionality of drag force to drag coefficient:

$$C_d = \frac{2F_d}{\rho v^2 A}$$

where:

C_d = Drag coefficient
F_d = Drag force
ρ = Fluid density
v = Speed of object with respect to the fluid
A = Object's projected area

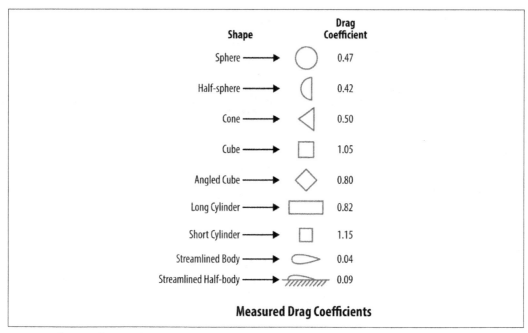

FIGURE 6-5. Drag coefficients of different shapes

Boundary Layer Separation

One factor affecting pressure drag is the separation of the boundary layer from a surface. Because the turbulent boundary layer is thin, it doesn't get pulled away from a surface as readily as a thick, laminar boundary layer. The longer adhesion of the turbulent boundary layer reduces the pressure drag because it decreases the size and strength of the deleterious wake turbulence.

In the case of a sphere, the boundary layer separation point for a turbulent boundary is about 120° from the impact point, versus only about 80° for the laminar boundary layer. What does this mean? With the laminar boundary layer, a large wake is produced with strong eddies. With the turbulent boundary layer, the wake is reduced. Therefore, the pressure drag is greatly decreased.

The dynamics of air separation and the vortexes produced by it have led to dramatic structural failure when the shedding of the boundary layer on either side of an object produces an oscillating motion. This vortex shedding phenomenon caused the failure of the Tacoma Narrows Bridge and the tower failure of an amusement park ride at Cedar Point in Sandusky, Ohio.

Mind Story: Golf Balls

Dimpled golf balls are an example of a surface that quickly creates a turbulent boundary layer in order to reduce pressure drag. The dimples make the surface rougher and trip up the boundary layer into turbulence. The stickier turbulent layer reduces eddy strength, decreases the pressure drag, and provides the ball with longer flight. In the case of the golf ball, the increase in friction drag produced by the dimples is more than offset by the decreased pressure drag.

This phenomenon is illustrated in the following image, which shows the relationship between wake and boundary layer separation.

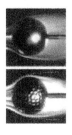

Wave Drag

When water is lifted up in the form of a wave, it works against gravity and therefore uses energy, as shown in Figure 6-6. This is wave drag and you see these in such examples as the bow and stern of boats and river pilings. The energy wasted in producing these waves is called wave-making drag and is only an issue in open flow systems like boats and drainage channels.

FIGURE 6-6. Water is lifted up as it flows into the piling

We should be humbled when studying drag by considering a commonly used equation for pressure drop produced in a pipe. This is an old, well-understood, and important equation. The formula is called the Hazen–Williams formula and it is used for filled, round pipe with turbulent flow. While it looks a bit nasty with all the nonintegral exponents, it is this very feature that illustrates the difference between theory and reality.

Pressure drop per foot in psi:

$$\Delta P = 4.524 \left(\frac{Q}{C}\right)^{1.85} \frac{1}{d^{4.87}}$$

where:

Q = flow rate in gallons per minute
C = pipe roughness coefficient (e.g., copper = 140)
d = inside diameter of pipe in inches

Closing Thoughts

Fluid mechanics seems to have the most bizarre names, which can take the joy out of describing how cookie dough and drinking fountains behave. You know most all fluid dynamics issues can be divided between those above and below Mach 1 and a Reynolds number of 2000. You know air can act nearly like a liquid and flow behavior changes dramatically with turbulent flow.

You know that drag is complex. Often, you want some surface roughness to cause a turbulent boundary layer to form, as this layer will be stickier and reduce pressure drag. You also know the surface properties of fluids produce resistance to flow and some fluids are non-Newtonian and do not act like water and air.

For a designer, fluid dynamics comes down to knowing that a slow design has less drag. If your design cannot go slowly, shape is usually more important than smoothness. You also need to understand the behavior of the fluid—some will change their behavior as they move faster.

If you are designing a boat, you want it to have nearly the same pressure at the front and back so you shape it like a fish, not a brick. You want smooth sides and gradual changes in direction. You know the static pressure on the underwater hull is huge. You know as the speed increases, shape becomes even more important.

If you are designing an automatic painting system, you know the first step is to understand how the paint behaves when it is pumped quickly versus slowly. You need to think how you are going to clean the paint out of the lines also!

CHAPTER 7

Heat Transfer

WE KNOW ALL ABOUT HEAT TRANSFER. EVERYONE HAS TOUCHED THAT HOT POT AND HAD OUR neurons awoken—ouch! We pay a lot of money to heat and cool buildings and houses. We buy fashionable coats to keep us warm and hats to keep the sun out of our eyes and off our heads (yes, and to keep us warm too!). We have touched sun-soaked objects and been sadly startled at how hot they were. We are fearful of boiling grease and hissing steam. We watch dogs pant and have no idea how small birds can survive frigid cold. We are happiest at a certain temperature and humidity but the world around us often wants us to be much hotter or colder.

In products and systems, heat is often created by friction and has to be removed. If you pull and squeeze a kneaded eraser you will be surprised how much heat builds up. Hopefully you have never touched a brake rotor after you park your car—the frictional heat produced by braking a car is enormous. Moreover, materials and our bodies behave differently when hot or cold. Studying heat transfer tells us why water evaporates at room temperature, why shiny seat belt buckles get so hot, why dogs pant, and why the SR 71 Blackbird airplane is painted black. It is for all of these reasons that the behavior of heat is important for a designer.

What Is Heat?

Heat is the movement of molecules—the faster molecules vibrate, the higher the measured temperature. The second law of thermodynamics tells us that this vibration energy wants to be shared; that is, heat flows to cooler objects. Heat transfer is accomplished by three mechanisms, although it is easier to simply think of five types of heat exchange: conduction, convection, radiation, infiltration, and internal heat.

The three heat transfer mechanisms—conduction, convection, and radiation—can be demonstrated by holding your hand at different positions relative to a flame. Sticking your finger in a flame produces conductive heat transfer from the flame to your finger (ouch!), placing your hand above the flame produces convective heat transfer from the flame to your hand,

and finally, holding your hand to the side of the flame (away from the rising hot gases) produces radiant heat transfer from the flame to your hand.

Conduction

Conduction heat transfer is produced by direct contact of two surfaces. The more insulated the material, the slower the heat transfers through it. In other words, aluminum with its high thermal conductivity will allow heat to transfer quickly through it, while air with its low thermal conductivity slows the rate of heat transfer.

For noncrystalline materials, conductivity principally relies on the movement of electrons for transferring energy. Therefore, materials with low thermal conductivity generally have low electrical resistance.

If constant heat is applied to one side of a block, a thermometer on the other side, and the whole arrangement is perfectly insulated (referred to as adiabatic), given enough time the temperature measured on the far side of the block will be the same as the applied temperature. However, if the measured temperature was plotted against time, the rapidity of the temperature rise would be directly proportional to the thermal conductivity. That is, a high conductivity would produce a rapid temperature rise. In practice, we rarely have adiabatic conditions; heat is always leaking out somewhere.

Thermal conductivity can be given per unit of material thickness, or as the total value per surface as seen in the thermal resistance (or R) values of house insulation. The R value gives the insulation of the total material per square foot. So a 1-inch (25-mm) thick foam panel has an R value of 3.5 per square foot (0.32 per square meter). If two panels were stacked on top of each other, the total R value would be 3.5 x 2 = 7. This is a convenient measurement because we can obtain R values for various types of building construction without having to add up the individual material conductivities and thicknesses. The U value is more commonly used in thermal load calculations. U is a measure of conductance and is equal to the inverse of R. Fourier's law describes how conductive heat transfer is equal to the multiplication of temperature difference, area, and the inverse of thermal resistance.

Temperature has other effects also. Most importantly for the designer is to recall that a material's expansion and contraction are directly related to temperature and the material's thermal expansion coefficient. The expansion and contraction of a material can tear apart materials and assemblies and was previously described in Chapter 3.

Convection

Convective heat transfer is produced by flowing liquids or gases. This is the principal method of heat transfer in many heat exchangers, including car radiators. The rate of convective heat transfer from a surface is related to the temperature difference and area (as it was with conduction) as well as the film conductance.

The film conductance, or convective heat transfer coefficient, is difficult to obtain. The best values are those based on experimental data, but the coefficient is proportional to the thermal conductivity of the fluid and inversely proportional to the thickness of the boundary layer. However, the general ranges of these coefficient values are as follows:

- Natural Convection — Air: 0.9–4.4 Btu/h ft² R (5–25 W/m²K)
- Forced Convection — Air: 1.8–35 Btu/h ft² R (10–200 W/m²K)
- Natural Convection — Water: 3.5–18 Btu/h ft² R (20–100 W/m²K)
- Forced Convection — Water: 8.8–1762 Btu/h ft² R (50–10000 W/m²K)

Natural convection arises from differing air density and the pressure it produces. Hot air is less dense than cold air and will rise due to the pressure differences. Forced convection means a mechanical system, such as a fan, drives the fluids.

Radiation

Radiation heat transfer is an interesting mechanism because it occurs by invisible electromagnetic radiation over vast distances without requiring a medium to travel through as with conduction and convection. Radiation heat transfer can be experienced when you feel the sun warming up your face on a cold, winter day, or the heat from a black road under the onslaught of sun. Interestingly, the propensity to absorb radiation is the same as that to emit radiation. If you put a piece of black paper and white paper in the sun, you will observe the black paper heats up faster. If you put the papers in the shade, you will see the black paper cools off faster.

Radiation will always be emitted when an object is hotter than its surroundings. The amount of radiation transferred from the hot surface to another surface is dependent on the temperature difference, emissivity of the surface (which will be presented later), and the orientation of the radiating surfaces with respect to one another. If a hot and a cold surface are parallel (therefore showing a lot of area to each other), they will transfer more heat by radiation than if they are askew. Almost no radiation transfer will occur if they are perpendicular.

NATURE OF ELECTROMAGNETIC RADIATION

Electromagnetic radiation is massless energy that radiates outward from a source. Unlike vibration and other mechanical wave phenomena, electromagnetic waves do not require a medium and can travel in a pure vacuum, such as outer space. In space, electromagnetic waves travel at "the speed of light" or 186,000 miles/s (300,000,000 m/s), while mechanical waves, used in ultrasonic thickness gauges, travel at only 2.8 miles/s (4.5 km/s) in structural steel and only a third of this velocity in rubber.

Visible light, X-rays, and radio waves are examples of electromagnetic radiation. Light is unique simply because our retina is sensitive to the range of frequencies known as visible light, with red light at the low frequency end and violet light at the high frequency end. Green lies in the middle band of visible light and the human eye can best discern the frequencies associated with green light.

Other frequencies outside the visible light band have important features. Infrared light warms skin while ultraviolet light causes sunburns. Microwaves heat water and X-rays let us see through solid objects.

A full description of electromagnetic radiation is seen through a description of its energy, wavelength, and frequency. High-frequency waves have a higher energy content than do low-frequency waves. Wavelength is an inverse function of frequency. Electromagnetic radiation has both wave and particle properties. Radiation behaves as particles, called photons, because they have defined length. That is, a light of one color will produce radiation of a certain frequency, but it will emit a barrage of photons.

The nerves in our skin respond to infrared frequencies and feel them as heat. However, our retina responds to the visual spectrum allowing us to see light. For temperatures below 1300°F (700°C), almost all the emissions are in the infrared spectrum and cannot be seen with the naked eye. The relationships between the frequency of electromagnetic radiation and some of its consequences are shown in Figure 7-1.

Where do colors come from then? Colors are a result of the reflection and absorption properties of a surface. Leaves are green because they absorb all the colors but green and reflect the green wavelengths. Black appears if the material absorbs all the visible wavelengths, and conversely, white appears if all the visible wavelength frequencies are reflected. However, we need to recognize that visible light is a very small portion of the spectrum and the surface properties of a material may be quite different at other frequencies.

The best example is snow, which is white and therefore is reflecting all the visible light frequencies. However, at the longer, infrared wavelengths, snow is an efficient absorber. Therefore, if we could see infrared radiation, snow would appear "black." Fortunately, at the high temperatures of incandescent lightbulbs and flames, radiation occurs in the visible light spectrum so lightbulbs illuminate things and we can usually see fires. Incandescent lights, for example, operate at about 4500°F (2500°C) and the sun is at approximately 10,000°F (5500°C).

FIGURE 7-1. Frequency of electromagnetic radiation and some of its consequences

RADIATION HEAT TRANSFER

Radiation properties are characterized by a balance of three material properties. These interconnected properties are transmission, reflection, and absorption. Transmission is a measure of how well radiation can transmit through a material. Transparent materials have high transmission while opaque materials have no transmission of light.

Reflection is a measure of how well a surface reflects—and therefore, rejects—heat energy. The nature of the reflected energy varies. The reflected radiation can be a specular reflection where the angle of reflected radiation is equal to the angle of incident irradiation, such as with a mirror. Alternatively, the reflections can be diffuse and uniformly distributed in all directions. *Albedo* is the climatological term used to describe the reflectivity of an object. It represents the ratio of scattered to incident radiation. Snow, for example, has a higher albedo at visual light frequencies than does grass, but they are nearly the same at infrared frequencies, as shown in Figure 7-2.

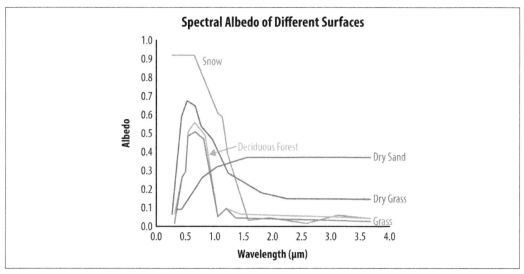

FIGURE 7-2. Albedo of materials changes with frequency, so that snow is very reflective at visible light (wavelengths from 0.4 to 0.7 nm) but not at infrared frequencies (wavelengths greater than 0.7 nm)

Absorption is a measure of how well a surface absorbs (as opposed to emits) heat energy. Reflection and absorption are surface properties and are affected principally by color and surface finish.

A unitless value between 0 and 1 is assigned to reflection, transmission, and absorption. These values indicate how much of the incoming radiation is divided among these phenomena. The summation of these phenomena is equal to 1 (or 100%):

absorption + reflection + transmission = 1

This means incoming energy is divided into three responses. It is divided in a manner dictated by material properties between absorption, reflection, and transmission. For example, in a nearly perfectly transparent material like glass, transmission is almost 100%. Therefore both reflection and absorption are equal to zero. If a material was perfectly absorptive, like a black brick, and did not reflect or transmit radiation, both reflection and transmission would be equal to zero and absorption would be equal to 100%. However, if a material was opaque but had some reflection, like a red brick, the transmission would be equal to 0 and therefore the reflection plus absorption is equal to 100%.

Absorption and reflection are properties of the surface of an object. That is, except for transmission, the only factor in considering the radiation properties of an object is its surface. Therefore, painting an object can change its radiation properties. For example, highly reflective roofs keep buildings cool, while highly absorptive roofs keep buildings warm. Besides painting roofs white (or other reflective materials), commercial buildings will sometimes use "spectral blue" windows that allow visible light to penetrate, but block the infrared light. This allows dwellers to enjoy the view without allowing the solar heating to drive up the cooling load.

Emission

Emission measures the ability of a surface to radiate. For a *gray body*, where reflected radiation is diffuse and the radiation properties are independent of wavelength, the emissivity equals the absorptivity. Emission is the ratio of energy radiated between one thing and a black body/object. A black body is the most efficient emitter and absorber of energy—both the emission and absorption are equal to one. That is, it absorbs 100% of the radiation that lands on it (there is no reflection and no transmission). Interestingly, human skin is a very good absorber and emitter with an emissivity equal to 95%. Figure 7-3 illustrates the response of an object to irradiation and internal heat.

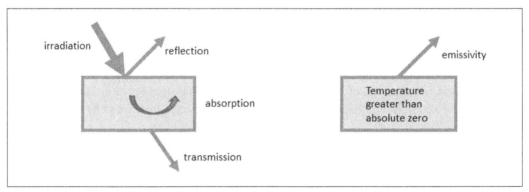

FIGURE 7-3. An object responds to incoming radiation by some combination of reflection, transmission, and absorption; an object emits heat as a function of its temperature and emissivity

To keep a surface cool, the material should have both high reflectivity and low absorptivity. It should also have high emissivity to get rid of any heat from its surface. However, we note for thermally opaque objects, where transmission equals zero, the reflection plus absorption equals 100%. That is, increasing one parameter decreases the other. Furthermore, absorption equals emission (Kirchoff's law of thermal radiation) so the interplay between reflectivity, absorption, and emission depend on the specifics of the problem being addressed. Table 7-1 provides the absorptivity and emissivity, at visible light frequencies, of various surfaces.

TABLE 7-1. Solar absorptivity of various surfaces

Surface material	Absorptivity
Aluminum	15%
White paint	25%
Concrete	60%
Colored paints	60%
Black polyethylene	94%
Black paint	94%

While radiation is emitted by everything with a temperature above absolute zero, the net amount of radiation between two objects is related to their temperature difference and area as well as the propensity to emit and absorb radiation.

If you put a small heated box inside another box at room temperature, both boxes emit radiation (because they are above absolute zero) but the net radiation will be outward from the hot box to the cool box. If they were at the same temperature but different emissivity and absorptivity, the inner box will receive radiation from the outer box plus reflections of its own radiation.

The temperature difference and surface properties explain why the ground loses less heat to a cloudy sky than a clear, dark sky. With a cloudy sky, the clouds are both warmer and have a higher reflection than the night sky. Therefore, the heat transfer is lower due to the decrease in temperature difference and the reflected energy radiated back from the clouds down to the earth's surface. Figure 7-4 provides an overview of the earth's response to solar irradiation.

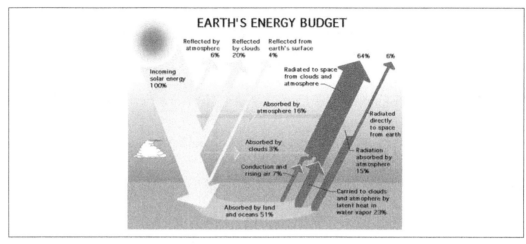

FIGURE 7-4. Earth's response to solar irradiation

Combined Effects with Radiation

Heat transfer is produced by the multiple mechanisms of conduction, convection, and radiation. However, radiation heat transfer is driven by temperature difference, area, and surface properties. Therefore, the conditions at the surface of an object are influenced by what is going on around the surface. Is the surface being continually heated or is it cooling off? Is the temperature much different than the surface it faces? Let's consider some of these questions.

RADIATION, CONDUCTION, AND SPECIFIC HEAT

The relationships between absorption and reflection raise the question of why, when your car sits in the sun, the seat belt buckle is much hotter to the touch than the seat belt webbing. You would think that due to the high reflection of the shiny buckle, it should not absorb much heat (absorption + reflection =100%). This is true. Because of its higher reflection, as shown in Figure 7-5, the buckle does not absorb as much heat as the webbing. However, the specific heat of the steel buckle is much lower than the webbing (about one-third the value). Understanding specific heat is the key to understanding the hot buckle problem.

Specific heat measures the increase in temperature per energy added. That is, specific heat measures the temperature response of a material to added heat. Therefore, the temperature of steel increases more quickly than does the temperature of the webbing for the amount of radiation they absorb.

Another factor in this story is that the steel buckles have much higher thermal conductivity than does the webbing. Consequently, when you touch the buckle, the heat travels very quickly from the hot buckle to your hand. The webbing will transfer the heat at its surface immediately but the heat travels slowly through the webbing so the cool spot you made with your hand stays cooler than the highly conductive steel seat buckle.

FIGURE 7-5. Watch out for hot metal seat buckles!

RADIATION AND CONVECTION

Heat exchangers, such as car radiators, both radiate and convect. However, convection is the dominant heat transfer mechanism in these cases. Some devices, including infrared heaters, radiant panels, and floors, are intended to heat by radiant heat transfer. While radiation is the dominant heat transfer mode, natural convection cycles occur and contribute to the heat transfer from these devices to the surroundings.

Sometimes heat transfer devices do not even try to get help from radiation. Electrical heat sinks, for example, rely almost entirely on convection to reject the heat of electrical components (see Figures 7-6 and 7-7). In this design, the large flat areas of the heat sink face each other. Because the two surfaces are at the same temperature, there is no temperature differential to drive radiation heat transfer. In reality, the radiation emissions are diffuse, so some of the radiation "escapes" the fin.

Note that fins must be parallel to gravity to take advantage of natural convection. If they are turned horizontal, convective flow and its accompanying heat transfer will be virtually eliminated.

FIGURE 7-6. Heat sinks that rely on natural convection except the one on the bottom left, which uses a liquid coolant

FIGURE 7-7. This heat exchanger uses a fan to increase convective heat transfer

Thermal Radiation Effects on People and Buildings

Cooling Load Factor

Radiation characteristics are expanded upon for comfort conditions by the use of the *cooling load factor* and *mean radiant temperature*. The cooling load factor accounts for the delay between irradiation and reradiation. For example, when a cool wall is irradiated by a warm person (that is, the person's infrared radiation hits the wall), the wall slowly heats up and then warms up the air.

At this point, we can introduce the notion of heat as it is used in heating and cooling design. Heat energy is considered in two categories: air or humidity. The first type of heat is called *sensible* because you can sense it. The second category of heat is called *latent*, which suggests it has energy associated with being in the liquid phase.

The cooling load factor refers only to sensible heat but people also produce latent heat. For example, when people enter a room, their moisture (from exhalation and perspiration) immediately enters the space. This is called latent heat gain and it does not have the time delay associated with sensible heat. The CLF adjustment can also be used for all internal loads, such as lighting and computers, to identify the sensible heat gain as a function of time.

Mean Radiant Temperature

The mean radiant temperature (MRT) is the average temperature felt when exposed to different surface temperatures. The MRT is a result of the amount of wall exposure and the wall temperature.

The simplified bioclimatic chart shows how infrared radiation affects human comfort. For example, a typical person would be comfortable if the air is cool and they are subject to infrared heating. The bioclimatic chart also shows how solar radiation, wind, and humidity affect human comfort.

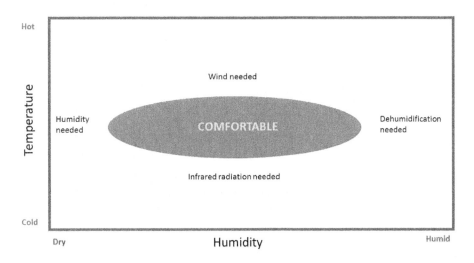

The American Society of Heating, Refrigerating and Air-Conditioning Engineers (ASHRAE) has an odd definition of thermal comfort, but it makes sense. The ASHRAE standard for comfort is a condition in which 80% of the people are not experiencing discomfort. Other issues relate to comfort beyond temperature, humidity, and draftiness. Carbon dioxide level and other indoor contaminants as well as noise affect comfort and safety.

Cooling Load Temperature Differential

Radiation is included with temperature differentials for surfaces exposed to the sun, like a roof. The term *cooling load temperature differential* (*CLTD*) is used to combine the heat from the outside air with the heat produced by the sun. The CLTD is dependent on the surface conditions (dark roofs absorb more heat) and the orientation with respect to the sun.

Other Heat Transfer Considerations

Heat doesn't just transfer; it can be created by phenomena such as chemical reactions, friction, and physical transfer of heated molecules from one place to another. The physical movement of matter is a common form of heat transfer and can be influenced by design features, just as convection can be encouraged by taking advantage of the buoyancy of heated gases. Heat energy can also be stored in helpful ways. Storing heat is vital for many designs and we will look at the interesting relationship between heat storage and temperature.

MASS TRANSFER

Mass transfer is an important source of heat loss and gain. This is not an energy transfer in the sense just described, but rather the actual exchange of molecules between spaces. Infiltration, unlike heat transfer that is driven by temperature differences, depends only on pressure differences. Air (or another fluid) is pushed into a new location. Infiltrated heat is not transferred between molecules as with conduction and convection, rather the molecules are actually exchanged. Infiltration is like dumping buckets of cold air into a warm room. Mass transfer describes evaporation also. In the case of evaporation, high-energy liquid water moves into the gas phase taking not only its mass away but also its energy. This is the process that people, dogs, and many other animals use to keep cool.

The Ideal System and Space

The ideal building would have a space wrapped in plastic so as to be airtight. In this room would be two holes. One would provide conditioned air, which means the air is low in carbon dioxide ("fresh" air) and appropriately heated, cooled, and humidity controlled (this is called *conditioned air*). The second hole would evacuate air so the room does not develop high positive pressure. Ideally, this second hole would also remove the stale (highest carbon dioxide) and hottest air, in the case of cooling, or the coolest air in the case of heating.

CAUSE OF INFILTRATION

The pressure difference that produces infiltration can be created by wind. This happens when a building develops high pressure on the side facing the wind and a low pressure downstream of the wind. The air then enters (infiltrates) through the windward cracks and exits (exfiltrates) through cracks on the downwind (or lee) side. The difference between the air coming in and the air going out establishes the building pressure.

A positive pressure makes doors hard to close but keeps contaminants out. Therefore, positive pressure is used for such applications as clean rooms and radon abatement. A negative pressure will slam doors shut but keeps the air from leaving a certain space. Therefore, negative pressure is used for such applications as kitchens and bathrooms to prevent malodorous air from escaping. Reference Chapter 6 for further exploration of this phenomenon.

In addition to wind, a pressure difference can also be produced by the buoyancy of warm air venting a building. The higher pressure of cold air forces warm air up, then viscosity and inertia entrain all the air in the upward movement called the *stack effect*. The stack effect increases with building height, as illustrated in Figure 7-8.

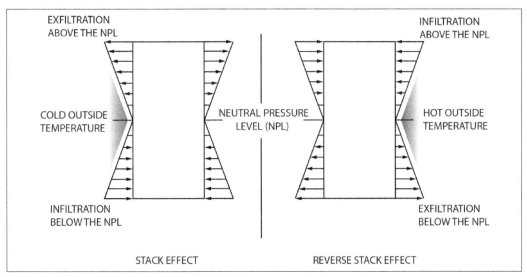

FIGURE 7-8. Rising warm air changes the pressure in a building so that it can draw air in at the bottom, called the "stack effect" (Image used with permission from ©ASHRAE www.ashrae.org HVAC Design Guide for Tall Commercial Buildings, Chapter 2, 2010.)

Pressure changes can also be produced by outside air. For example, wind blowing on a building will produce a higher pressure that can promote infiltration. An example of pressure change from wind being used advantageously can be seen in termite nests, which have a large central chimney that produces a cool nest, as indicated in Figure 7-9.

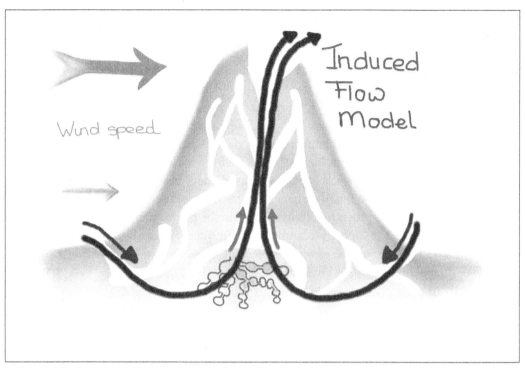

FIGURE 7-9. Wind blowing over the nest draws air out of the termite nest

THERMAL STORAGE

The energy stored is a function of the specific heat, temperature change, and mass. We had considered specific heat in the earlier discussion of the hot seat belt buckle.

Specific heat is a measure of heat storage per degree temperature change of a material. For example, the specific heat of water is over 10 times that of copper. Therefore, it takes a lot more heat to elevate the temperature of water 1° than it does copper of equal mass. Because the temperature change is also a function of mass, it can take more heat to heat up a huge mass of copper than a small mass of water. Figure 7-10 shows the disconnect and wide range of variability between the specific heat and thermal conductivity.

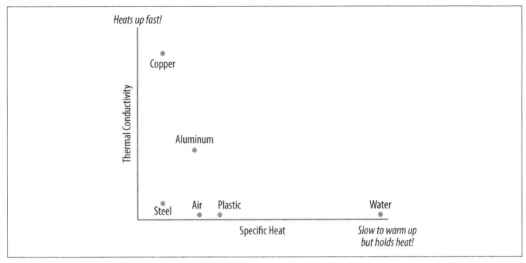

FIGURE 7-10. Plot of thermal properties for common materials

INTERNAL SOURCES OF HEAT

Heat is generated by such things as people and electronics. As discussed previously, heat is produced in two forms: sensible and latent. Sensible heat can be "sensed" and measured with a thermometer, while latent heat is the addition of moisture to the air. Lights, for example, produce only sensible heat, while people produce both sensible and latent heat. Well-insulated commercial buildings, such as offices and stores, have such high internal loads that they are actually self-heating until the temperature falls below around 35°F (2°C).

Mind Story: SR 71 Blackbird

Why is the SR 71 Blackbird painted black when the paint adds 100 lbs (45 kg) of weight?

The black paint increases emissivity and therefore cools the airplane. This airplane travels over 2200 mph (3540 km/h), which makes the air molecules compress and scream across the surfaces pro-

OTHER HEAT TRANSFER CONSIDERATIONS | 135

ducing a lot of friction. Because of the frictional heating at high velocity, the airplane gets really hot, over 600 °F (315 °C). The only way to get rid of heat in a vacuum (or the near vacuum of high-altitude reconnaissance flight) is through radiation. Therefore, the surface properties of the airplane are such that they maximize the emission of the heat radiation. Yes, the black color will absorb more solar radiation than another color, but this is small compared to the enormous frictional heating that must be shed. There were other reasons for the paint—it made it more stealthy. The paint contained ferrite particles that absorbed radar signals.

Thermal Calculations, by the Numbers

Thermal resistance factor

$$U = 1/R$$
or
$$U = k/L$$

where:

R = insulators R-Value
k = thermal conductivity (per unit thickness)
L = material thickness

Conduction

$$Q = UA\Delta T$$

where:

U = conduction resistance. This is equal to $1/R$
A = Area
ΔT = Temperature Differential

 When considering combined effects with radiation in a building, the ΔT is replaced by a CLTD value.

Convection

$$Q = hA (T_s - T_{inf})$$

where:

h = film conductance
T_s = Temperature at surface
T_{inf} = Temperate at infinite distance from surface (ambient)

Frequency and wavelength

$$\lambda = c/f$$

where:

λ = wavelength
c = speed of electromagnetic radiation
f = frequency

Radiation

$$Q = \varepsilon A S (T_{s1}^4 - T_{s2}^4)$$

where:

ε = emissivity
A = surface area emissivity
S = Stefan–Boltzmann constant = $0.1714 \; 10^{-8}$ Btu/h ft^{2}R^4 ($5.67 \; 10^{-8}$ W/m^{2}K^4)
T_{s1} = temperature of surface 1
T_{s2} = temperature of surface 2
Note: Q is proportional to solid angle at $1/r^2$

Heat produced by people

$$Q_{people} = (N)(SHG)(CLF)$$

where:

N = number of people
SHG = Sensible Heat Gain per person
CLF = Cooling Load Factor (typically starts at 0.6 to 0.75 for the first hour)

Specific Heat

$Q = \rho \, c \, V \, \Delta T$

where:

ρ = Density
c = Specific Heat
V = volume
ΔT = Temperature Differential

Closing Thoughts

Heat transfer is one of the more intuitive subjects we have discussed. Heat transmits by contact with a material or via radiation. You know heat can be produced internally by such things as chemical reactions, living organisms, and friction. If you are designing an electrical circuit that will reside inside a cabinet, the best way to get rid of the heat is through a cooling liquid that directly contacts the component within the circuit that is producing the most heat. The transfer is increased by having very cold coolant (giving a high temperature difference), a lot of contact surface area, and thermally conductive materials. If you cannot route a liquid coolant, convection will help. You know good convective cooling needs cool, high-velocity fluid and a lot of surface area. If this is not an option, you know you can rely on a natural convection—that is, taking advantage of the movement of warm air to run across your heat exchanger. However, you know cooling fins need to be oriented up and down so they are aligned with the air flow. If you have to rely on radiation, you remember the hot component has to have something cool to radiate to and requires lots of area as well as a high-emissivity surface finish.

If the frame that holds your automatic pancake flipping invention is getting too hot, you might want to check the rotating surfaces to see if you have a lot of frictional heating. If that doesn't fix the problem, maybe the device is receiving too much conductive heat from the pan and you need some thermally insulating materials. Perhaps your frame is getting too much radiant heat from the stove and you need to reduce the exposed area and paint it a shiny, non-absorptive color.

CHAPTER 8

Human Factors

DESIGNERS USUALLY DESIGN FOR PEOPLE. WHILE THE COMPLEXITY OF THE MIND IS DIFFICULT to unpack, the body is more straightforward. Humans are mechanical linkages, muscles, and sensory systems—all of which are driven by both our stomachs and our minds.

In this chapter, we cover eclectic topics related to human factors. We consider how people use things and how they move their bodies. We evaluate human dimensions and how to make shapes fit people comfortably. We also briefly review the notion of comfort and how people interface with the world. This chapter is a bookend to the first chapters where we looked at nontechnical issues. When considering perception and the mind, we often rely on guidance derived from experience rather than the march of physical science presented in most of the preceding chapters.

Ergonomics

Ergonomics is the science of relating design to the human body. The goal of ergonomic design is to allow a device, system, or information interface to easily connect with a human. Ergonomic designs are more comfortable to use and reduce injuries and accidents.

The concepts behind ergonomics include a study of human dimensions (anthropometry), human movement (kinesiology), user psychology, interaction design, and environmental design. The idea of comfort is very subjective and involves every aspect of the user environment. A chair squeak might be more irritating than lack of lumbar support for some. An automobile's poor cup holder placement may be more irritating than lack of intuitive radio controls for others. Comfort balances concerns for form and dimensions, tactile responses, sounds, and mapping. Recall in Chapter 7 we saw that that environmental comfort was dependent on factors beyond simple temperature, such as humidity, draftiness, air quality, and noise.

Fortunately, ergonomic design is surprisingly straightforward—simply design human interfaces that put the body in a neutral position. Injuries or discomfort increase when the hand, foot, arm, and back. are compelled to operate outside their neutral position.

Ergonomic design is based on data, not intuition. You must specifically avoid the mistake of designing for yourself and assuming it will be satisfactory for everyone else. Moreover, designers should not assume a design for the average person is satisfactory for those at the outer ranges. Anthropometric and kinesiologic data are readily available for a wide array of demographics and provide guidance for developing designs that keep the body aligned in the neutral position and offer the appropriate range of customization required.

Handle Design

We all know how we want handles designed. They need to be rounded, slightly textured, and support the surfaces for palm and finger as appropriate. In addition to these commonsense approaches, data provides the following guidance. Remember, they are rules of thumb only:

Handle Diameter
- 1.2 in.–2 in. (30 mm–50 mm) for power grip
- 0.32 in.–0.63 in. (8 mm–16 mm) for fine work

Handle Length
- 3.9 in.–4.7 in. (100 mm–120 mm) long for power grip

Hand and Finger Clearance
- 6.5 in. x 1.6 in. (115 mm x 50 mm) for handles (e.g., suitcases)
- 1.4 in. (35 mm) for finger or thumb affordance
- 1 in. (25 mm) radii where hand might engage

Anthropometry

Objects designed for people different than ourselves require data because we can't simply create a design and "see if we like it" or if we are comfortable with it. Anthropometric data provides detailed dimensions on the human body and considers range of motion. This data is obtained by measuring devices that capture dimensional and reach information. Reach information can consider factors like strength reduction as a function of distance from torso, dexterity requirements, and biomechanical restraints on motion. Anthropometric data also includes what is in our line of sight and helps us locate instrumentation and human control inputs. Anthropometric data is obtained empirically and is a tough job because the data varies with gender, age, and ethnicity.

[**Statistics**

Approaching most data requires an understanding of the statistical nature of the world around us. People aren't the same. We have all sorts of different lengths and weights so anthropomorphic data is given as an average value and a standard deviation (referred to as σ or SD) or as a percentile of the population. The standard deviation indicates the range of the data—that is, how close or far the spread of data is throughout the average. A high standard deviation means the data goes high and low around the average. This high standard deviation makes the average value less helpful in design because there exists a wide variation. Conversely, a low standard deviation means the range of data does not vary far from the average.]

Anthropomorphic data is widely available. Figure 8-1 represents neutral body posture based on NASA data and Figure 8-2 shows the body proportions as a function of height. A source of more precise data is available in the Anthropometric Data Explorer feature at *www.openlab.psu.edu*.

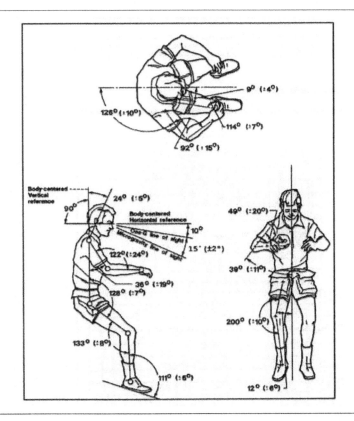

FIGURE 8-1. Neutral position

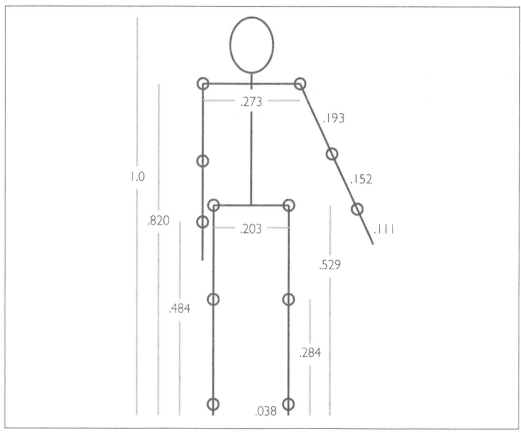

FIGURE 8-2. Body proportions as multiples of a man's height (image courtesy of Parkinson, MB. Proportionality constants. www.dfhv.org)

Some correction factors that are helpful when dealing with raw anthropomorphic data include adding 5% to values to account for light clothing. People can stretch 10% further than anthropometric data suggests by twisting their torso and extending their reach.

Designing with Anthropometric Data

You should use the anthropometric data appropriate for your age, sex, and ethnic group. Typically, you design your maximum dimensions (for adults within a specific ethnic group) for all but the largest 5% of men and your minimum dimensions for all but the smallest 5% of women.

Another best practice is to design for the neutral position and/or the middle of the range of motion. The effectiveness of your design needs to be verified empirically. Remember that ergonomic design is based on data, not intuition. You need to specifically avoid the mistakes of designing for yourself and assuming it will be satisfactory for everyone else. Moreover, you should not assume a design for the average person is satisfactory for those at the outer ranges.

A simple way to get a sense of a design's suitability for someone different than you is to quickly build a full-scale anthropomorphic figure using wood or PVC piping. It only takes about 15 minutes. Plus you can put a balloon for a head and you're done. You can also use wire to fill out the body if you want.

Another modeling technique is to locate the eye level and set up a transit at this point. From this position, you can measure sight lines. Using basic geometry, you can determine the suitability of the sight lines or reproduce a cockpit, for example, based on angle and distance measurements.

SEAT DESIGN EXAMPLE

While many texts and websites provide anthropometric data, let's consider a couple of designs that are closely tied to anthropometry.

If you were to design a chair for the mass market, you would first find three data points in order to get the correct starting point for seat height, depth, and width. These data points are (1) the right angle seated (popliteal) height, (2) the buttock to popliteal length, and (3) the hip breadth. Popliteal refers to the area behind the knee and is a word intended to scare you.

Both seat height and depth should be low enough to accommodate the smallest 5% of women. If the seat height is adjustable, you can accommodate more sizes comfortably. Most people want to adjust seat height by plus or minus 10% of their popliteal height. The width of the seat would accommodate 95% of men and women. If you compare the biggest men and women's hip breadth, women's will be a little larger so the seat width dimension is dictated by women's anthropometric data. However, to make the seat practical, you have to consider the bulk of clothing, and for those not driving a Formula 1 racecar in a bolstered race seat, extra space for the hips. Therefore, for the seat width, you might add a clothing allowance and an oversize factor. However, if the seat does not have constraints on the side, you can make the seat smaller than the maximum hip breadth (usually about 1 in. or 25 mm smaller). This doesn't sound very creative but it is real data for real people and is a "hard point" for dimensional guidance. If the seat reclines, another set of concerns arise. The angle of recline, not just the back angle, is important in distributing loads. Ergonomic design requires empirical feedback, not just a take-it-or-leave-it approach of hard numbers and scientific imperialism.

Note that anthropometric data is connected with ethnicity, gender, and age. Also, this example considered the percentage of people you are trying to accommodate. The 5% smallest and 95% largest is a rule of thumb but obviously this would be unsuitable for children or obese populations.

> **Designing for Ergonomics**
>
> - Use anthropometric data. Typical size range for all adults within an ethnic group:
>   — Maximum: 95% men
>   — Minimum: 5% women
> - Joint position should be neutral, typically midpoint of their range of motion.
> - Work should be done by largest appropriate muscle groups.
> - Avoid overexertion of connective tissues (muscles, tendons, tendon sheaths) in wrist.
> - Avoid compression of the median nerve in the wrist's carpal tunnel and use a natural grasp angle of 60°–70°, limit movement to 15°.
> - When lifting is required, avoid wrist pronation (palm down) and supination (palm up). Also ensure that the upper arms can be close to body and elbows operating around 90°–100°.

Kinesiology

Kinesiology is the scientific study of human movement and its application ranges from orthopedics to exercise equipment. Kinesiology synthesizes biochemistry, biomechanics, motor control, physiology, and psychology in studying human movement as it relates to exercise and health.

Human motion can be described geometrically by examining the limits on rotation of various body parts. However, at the extremes of motion, the body part may have less strength and dexterity. In addition, motion at extreme ends of human limits can cause injuries and increase the time requirements for a task.

You experience this if you attempt to use fine motor skills in your hands while they are folded behind your back. You can feel the muscles complaining about this type of work. Activities that require you to look all over the place also slow down your ability to acquire information.

WALKING AND RUNNING

Walking is the normal mode of motion rather than crawling or running because it is most efficient for humans. Walking is a complex phenomenon in which biological energy is converted to motion while maintaining balance, absorbing impact, and joint contact.

Walking has to overcome the large friction associated with moving on rough ground. In walking, the process of lifting your foot reduces friction but picking up the foot requires overcoming gravity by going in the wrong direction (up versus horizontal). And now you have introduced a balance problem, which is solved by delicately controlling core muscles: lifting up the foot, swinging leg, heel strike, and foot strike.

The process of shortening one side of your body to reduce drag force involves lifting up your pelvis, bending the knee, and then twisting the foot. The hips go through a figure-eight motion as you walk and your head bobs up and down in balanced response. The lifting of the foot takes energy, but you have stored this as potential energy that is recovered as kinetic energy on the downward foot strike. This complex motion is also governed by the pendulum effect of your leg, which has a natural frequency, like a metronome. Cross-country skiing and elliptical trainers reduce the friction problem by allowing the feet to slide back and forth. This sliding motion is therefore more efficient than lifting the feet in walking.

Energy requirement increases exponentially with speed. So there comes a point when running is more efficient than walking (typically about 5 mph or 8 km/hr). Running is different from walking because you don't always have one foot on the ground as with walking. As speed increases, inertia increases and provides stability. This reduces the balance requirements that compel us to have at least one foot on the ground when walking.

Mapping

Mapping describes the relationship between visual cues and function, such as scissor handle movement mirroring the cutting blade action. The mapping of a product or system is an important design concept. In products and systems that are vital to safety, it is critical to make the mapping as clear as possible. Deviating from this approach requires thoroughly training the user. Consequently, traditional mapping needs to be taken into account so that a new design is approachable by those who have experience with an old design. Affordances and constraints should also be introduced to design to accommodate or constrain how a device is used. An example of this is scissor handle holes that are sized to handle either the fingers or thumb.

Scissors, as shown in Figure 8-3, are one of the best examples of intuitive use. They have excellent mapping as well as inviting affordances for your fingers.

FIGURE 8-3. Scissors are intuitive and provide appropriate affordances

Some designs that also cut paper are more difficult to figure out (Figures 8-4 and 8-5).

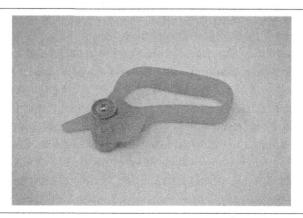

FIGURE 8-4. What is this thing?

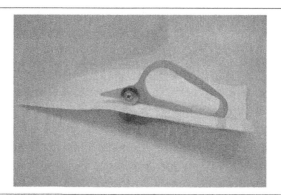

FIGURE 8-5. It is also used for cutting paper! Some instruction required.

Human Injuries Related to Design

Doing work, using products, or even sleeping funny can cause injury. The body does not like to work with heavy weights, fine motor skills, twisted limbs, and long reaches at the same time. If high strength is required, the activity should be done close to the torso; if fine motor skills are needed, the strength requirements should be low; if a long reach or complex twists are required, both the strength requirements and fine motor skill requirements should be low. Poor design both of products and activities can lead to a large category of musculoskeletal disorders (MSD). These disorders describe injury to the soft tissues that cause pain, numbness, tingling, and muscle weakness.

In addition to carpal tunnel syndrome, described previously as the compression of the median nerve in the carpal tunnel, other nerve injuries include thoracic outlet syndrome, radial tunnel syndrome, and cubital tunnel syndrome.

The thoracic outlet is a crowded space between your collarbone and first rib. If the shoulder muscles can't keep the collarbone in place, it will start compressing the nerves, muscles, and blood vessels underneath, producing the category of thoracic outlet syndromes. This syndrome can be caused by poor head position or slumped posture.

Radial tunnel syndrome is a compressed radial nerve on the outside of the elbow resulting from wrist and finger extensions. Cubital tunnel syndrome is a compression of a nerve inside of the elbow arising from bending the elbow or leaning on your elbow. This syndrome is caused by pressure on the ulnar nerve that lies close to the skin's surface under the "funny bone."

Basic Kinesiology Vocabulary

The body has three imaginary places that allow clear expression: sagittal, transverse, and frontal. The sagittal plane splits the body down the "midline," or middle, dividing left from right. The coronal or frontal plane is the division between the front and the back of the body. The transverse plane divides superior part of the body (toward the top) from inferior (toward the bottom).

Pronate
    Turn the palms up or the feet inward.

Supinate
    Turn the palms down or the feet outward.

Abduction
    To move away from the middle of the body.

Adduction
    To move toward the middle of the body.

Extension
    To straighten out a joint.

Flexion
    To bend a joint.

Understanding basic anatomy, especially that of the hand, is helpful in design. The hand has muscles that control fine movements, and external muscles in the arm that propel powerful movements. The muscles are attached to the hand with tendons. The tendons are sheathed in a protective and lubricating membrane that allows them to move through bony structures.

The tendons that connect the exterior muscles in the arm to the hand bones run through a small opening in the wrist's carpal bones. The tendons share this carpal tunnel with nerves and this area is vulnerable to injuries related to inflammation. Carpal tunnel syndrome is produced by a compression of median nerves in the carpal tunnel due to inflammation.

The bones in the hand constrain how the hand moves. The bone protrusions act like levers, which along with the geometry of the muscles and tendons dictate the fingers' force and motion. Hand anatomy uses a logical repeating nomenclature for the five digits and these can be helpful to memorize if you work on designs involving the hand. Moreover, hands comprise one-quarter of human bones (and the foot bones are very similar), so knowing these two structures allows you to readily memorize half of all the bones in the human body.

Biomimicry

Biomimicry means copying nature's models in human design. This is seen when you compare the streamlined shape of birds and fish with airplanes and boats. These shapes reduce pressure drag and are more efficient than other shapes. Often, biomimicry is helpful when looking at the shape of the surfaces of living organisms, such as the hooks of burrs, which were copied to create Velcro.

Other natural surface shapes are also employed in design. For example, drag through the water can be reduced by copying the shapes in sharkskin and the bumps on the leading edge of whale flippers. Both of these features have been copied in swimsuits.

Biomimicry can complement sustainability and is often pursued in terms of manufacturing and extraction techniques from self-assembly to carbon dioxide conversion to microbial metal extraction. Biomimicry is also pursued in other fields, such as control systems, sensors, logic, adhesives, and high-temperature encapsulation. Observe nature and learn great things!

Interaction Design

Interaction design focuses on behavior more than products and is usually related to virtual environments. This discipline manages images, sounds, and touch associated with humans contending with data or devices. The world "behind the glass" of mobile devices is huge but the human interface is challenged by the small size and the mass of data. Interacting with machines has gone from levers to buttons to keyboards to touchscreens—and there is still a lot of work to do to make this efficient.

Many guidelines have been promulgated that connect virtual designs with real people. For example, people prefer to recognize rather than recall a word so that user prompts should offer potential options rather than emptiness. Also, people like feedback on computer processing so we know that something is actually happening and we don't need to press a button again. Standardized formats that provide repetition and familiarity, connection between the real world and computer icons helps us navigate. No one wants to read a user manual, so good interaction design is necessary for providing intuitive navigation of the virtual world.

The key to good interaction design is rooted in your ability to empathize with the user. Typically, you become an expert and tend to reflect potential system capabilities with a detailed array of options. The casual user wants easy-to-approach interactions and immediate results, such as seen with a power button or volume control knob. This simple first interaction moves into the more complex world of engagement where users can use a system in such a manner that their work is the only thing on their minds and not the interface.

The 80/20 rule (Pareto principle) states that 20% of things are vital and 80% are trivial. Because Pareto was an economist, he noted that 20% of the people owned 80% of the wealth. This principle can be extended into interaction design by considering that users will employ 20% of your design features 80% of the time. These 20% need to be easy and obvious, in addition to providing immediate feedback, much like the mapping described previously.

Error prevention is another key concern in developing interactions. Anticipating errors (such as misspelled words or the pushing of nearby buttons) allows you to apply beautiful human reasoning to the whole design. Automatically correcting likely errors is helpful, as is disabling features that aren't needed by the user and could mistakenly be used.

Fitt's law states that it becomes progressively more time consuming to click or point on an object the further away or smaller it is. This is a big issue in the virtual world where more information is compressed onto a smaller viewing area so pop-up windows and cursor dynamics become part of the interaction experience.

Expert Systems and Artificial Neural Networks

The benefit of expert systems and artificial neural networks is their ability to model human thought in a manner that can be processed on a computer and recalled by a nonexpert user. However, sharing human knowledge and expertise is a difficult task. People use sophisticated and intricate thought processes to solve problems, recall information, and make decisions. Although expert systems are an excellent technology to save and disseminate knowledge, transferring that knowledge from the domain expert to the computer is difficult. Acquiring knowledge for an expert system is the art of structuring human instincts, experience, "rules of thumb," guesswork, and all the other words that never can quite explain the human thought process.

Expert systems have been available for many years to aid in the diagnostics of machinery, systems, and even the human body. Expert systems have had a frustrating history as they try to emulate human thinking. They work reliably on focused problems but generally falter when asked to contend with highly complex systems. An expert system is a collection of heuristics or "rules of thumb" that are assembled to aid a user in diagnostics and problem solving. Expert systems use fuzzy logic to make decisions. That is, the confidence of an output is quantified and this value can be handled separately from the output. The confidence output is then used in determining the best, final decision of the expert system. If multiple inputs seem to suggest a certain output but all of these inputs are of suspect value, then another more likely output is generated. The combination of outputs and the confidence associated with them adds intelligence to the system. In this way, the "degree of truth" of information can be managed.

Artificial neural networks use a large number of processors with each artificial neuron dedicated to a specific task. The neural networks organize the links between inputs, outputs, and hidden intermediate layers of decision making. Sensory or database information is fed through this network with each neuron processing the data independently and progressing its results through the network. Generally, a feed-forward approach is used where information flows from the input neurons, through the intermediate neurons, and finally to the output without a feedback mechanism.

Neural networks differ from expert systems in their ability to remember information and adapt to changes in the incoming data. That is, they have the ability to learn rules themselves based on data. For example, the neural network can be provided a large set of data and it will automatically find all correlation between the data. These systems are widely used for predicting financial trends and connecting social networking or other big data to suggest everything from purchasing decisions to criminality.

Closing Thoughts

Human factors are a huge category. When people interface with your design, there are many things to consider. The physical and mental fit are vital. Our bodies are less adaptable than our minds, so ergonomics was presented with the simple notion that good designs keep the body aligned in the neutral position and offer the appropriate range of customization required. Anthropometric data can be fed into your design detailing to ensure good fit.

The mental connection with design is tough. Interaction design is an expansive and dynamic field. However, user psychology and all the nontechnical issues discussed in Chapter 2 are vital. Many design tools help communicate human factors such as experience maps, mind maps, affinity diagrams, and wireframes. Artificial intelligence, in its various forms, guides users in a rapidly evolving format. However, designers and inventors need to empathize with their target market through various means of research and ethnography. Synthesizing these findings can motivate designs that might violate strictly technically derived approaches. How do people really use your design? What are their pain points? What do people want that they don't yet have available? These are some of the fundamental questions you need to consider.

CHAPTER 9

Sustainability in Design

GRASS AND WEEDS ARE FORMS OF BIOMASS AND WHEN BURNED PROVIDE A RENEWABLE FORM of heat energy. One proponent of burning willow weeds claimed this fuel produces a tiny amount of ash. Not really. Even though only a small amount of ash was collected in his heating system, the amount of ash generated was unknown because the combustion system had very ineffective pollution control equipment. It only collected the largest particulates. Therefore, he only saw a little ash in the boiler. Most of the ash went unseen up the exhaust stack and is thought to never have existed. However, it did exist and went out the stack as tiny particulates.

Beyond misunderstanding how fine ash gets produced (and gets deeply inhaled in the lungs), we need to consider other aspects of this renewable fuel source. Do these fuels produce air toxics such as dioxin or mercury? Do they produce nitrous or sulfur oxides? What chemical might be attached to the ash particles? What effect does growing these plants have on soil, or fertilizers on water nitrate levels? What heat source is it displacing, and is this really an improvement in terms of air, water, soil quality? Some level of technical sophistication is required to discuss sustainability fairly.

The word sustainability is commonly seen now but it has been around in different forms for ages. Sustainable design is a holistic approach whereby you consider all activities involved in getting a product into the end user's hands and what that user does with the product at the end of its useful life. Sustainable design compels you to recognize the source of materials and labor and the concomitant environmental and human effects. It is not very complicated—it is just sensitizing yourself to these matters. Sustainability is considered in a process of life cycle assessment described by the ISO 14040 standard.

Our material culture has changed due to the globalization and professionalization of design. Design rarely reflects the specific environment we live in. We don't cut down some oak trees in our forest to make a dining room table while keeping in mind how many trees we will need for heating this winter. We don't really need to balance what we consume with what we will need in the future in this direct manner—we just buy stuff. After we buy it at a certain acceptable price, we use it, then eventually throw it in the garbage.

These are three separate issues to ponder from a sustainability standpoint: backstory, usage, and disposal.

Backstory

What is your design made from? Sometimes the backstory is the only story; consider Marcel Duchamp's *Fountain*—an old urinal with a small signature. He declared it "art" and it is now considered one of the most important works of modern art.

A good concept to carry is the notion of a material or product containing *embodied energy*. That is, energy involved with the extraction, processing, packaging, transportation, storage, and administration. Social costs are another element to sustainability. What deleterious effect does a material make on the ecosystem or people? This study of sustainability can tend to be preachy. However, it is appropriate to think of the impact materials make on a broader scale than simply the purchase price. Most of the effects are good: employment and useful products. Some of the effects can be bad, such as air pollution and child endangerment. These are examples of impacts that do not appear in the price of materials.

Embodied energy constituents:

- Mining
- Processing
- Packaging
- Transport
- Storage
- Administration

Even recycling has embodied energy. Consider this simplified listing of steps involved with recycling a plastic bottle:

1. Transported to a collection facility
2. Inspected for contaminants
3. Washed
4. Chopped into flakes
5. Flakes are dried and melted into plastic lava
6. Lava is filtered for impurities and formed into strands
7. Strands are cooled in water

8. Strands are finally chopped into pellets that can go to market

But this is just looking at recycling; let's consider the embodied energy involved with a brass gear. Brass is a copper alloy. Most copper comes from deep mines in Chile where the ore is shipped to a smelter. Then the ore is ground up by giant machines with equally giant electric motors. The copper is melted with high-temperature furnaces, then cooled and sent to be melted again and mixed with other material to make alloys. The alloy is then formed into bar stock and machined into a gear.

The production of a gear leaves most of the bar stock from which it started as chips. In addition, any production mistakes end up as scrap also. The scrap can be recycled, which is good but it has machine oil on it that needs to be cleaned off. Where does that wash go? How is the air pollution from the melting operations and power plants supplying electricity for the electric motors handled? What becomes of the mine runoff? The answers depend on national laws. All these steps involve energy of some form and the production of air and water pollution as well as human labor.

Another part of the backstory is storage and transportation. What is involved with transporting all the materials to final production and then moving the final product to the consumer? Are items stored in climate-controlled warehouses? Are bribes paid to expedite transport? Or even worse, are undemocratic governments sustained by the payment for the material?

Packaging is a visible but small part of sustainability. Styrofoam peanuts and overwrapped CDs of the past are examples of sustainability issues related to packaging. Packaging is vital for some companies, especially in the cosmetics business. The unpacking and enjoyment of the perfume bottle is part of the experience, as too is the unwrapping of the long-awaited Apple product. The packaging is exquisite, but you could argue it is not efficient. Some may remember the foam-encased McDonald's hamburgers—great packaging, but an environmental mess, and now we are generally back to paper wrapping.

Administrative costs are another example of embodied energy. For example, real ecological impact is produced by any administrative elements, such as inspections and certifications, which are required from some element of the material, production, transportation, storage, and distribution. These activities might destroy material; repair or repackaging could be required; all these employees probably use fuel to get to their jobs, which are in air-conditioned buildings; and so on. I remember an inspector who had to verify the impact resistance of a new product. He drove many hours to the test site, sat and drank coffee while watching the test, and then drove home. He ensured the test was conducted honestly, but there was an ecological impact as well as opportunity cost to his inspection. This identification of embodied energy does impart a value (I think we all want our food carefully inspected!), but it is a matter of being honest with the broad ecological impact of making things.

Sustainable design versus unsustainable design is not always clear. For example, photovoltaic (PV) cells produce electricity from sunlight. Wonderful! However, they are not mined from a PV mine, they are manufactured using extracted materials and giant machines. Electrical-grade silicon is made by heating quartz and sand with carbon forms such as coal that produces low-grade silicon and carbon dioxide. The low-grade silicon is reacted with an acid, heated, and then cut to produce the electrical-grade silicon used in PV cells. We must ask how PV cells are made and what else goes into them before they end up producing electricity. Some questions are even philosophical; for example, should food be burned as fuel?

Usage

One of the biggest questions here is this: how is the design used and misused and where does it derive its energy? A rubber band can hold papers together or can be fabricated into a zip gun. Fertilizer can help crops grow or be used to make a bomb. Nefarious uses aren't the only concern for a designer. One of the more artful insights required is anticipating the loads and environments a design might encounter in normal usage, which was discussed previously in Chapter 3. These are not always obvious.

Energy source is another consideration. If a product uses rechargeable batteries, it is getting much of its power from coal-fired utilities (or nuclear, in some countries)? If it uses batteries, where do they come from and what will become of them when they are disposed?

Disposal

What happens when you are finished with a product and its packaging? Do you throw the packaging away as a mass of sharp-edged plastic with foam peanuts clinging along? Do you throw the worn-out design in the garbage some time later? Sustainable design, in this context, means considering if recyclable materials can be easily removed before throwing something away. Are people likely to separate recyclable from unrecyclable? Do manufacturing techniques such as insert molding prevent even valuable copper or aluminum from being separated from the molded material around it?

Even more important than recycling is the proper separation and disposal of potentially hazardous materials. Can these materials be easily removed and would people want to bother? What happens to your design in a landfill? Does it decompose, leach out nasty stuff, or just sit there?

E-waste is defined as anything with a battery or a cord. The mass of E-waste is increasing rapidly. Every year, the United States currently generates about 49 pounds (22 kg) per person, while the global average is 15 pounds (7 kg) per person. This includes the 2-year replacement cycles for phones. E-waste can have materials that are harmful to humans and the environment if not separated and treated.

Once we are born, we are impacting the environment, and all the things we design will also have some impact on people and the environment. The notion of sustainable design is recognizing the whole cycle of production and consumption and then striving to maximize the good stuff while minimizing the bad stuff.

Milton Friedman offered an economic discussion about the making of a pencil and how market forces align the makers of graphite, rubber, wood, glue, and paint into a final product that is sold cheaply worldwide. This market-driven, transnational cooperation can lead to high efficiency and low cost, much like Adam Smith's famous description of an efficient, 18th-century pin factory. However, these classic models of the complex nature of economic forces in design and manufacturing do not consider sustainability directly. This is because sustainability is difficult to measure monetarily and the rules and ethics associated with it change over time.

Recycled Fraction

The fraction recycled is a measure of the proportion of a material in use in products that can economically be recycled.

- Materials that can be remelted (like metals or thermoplastics) or shredded (like paper and wood) are most easily recycled.
- Some materials are difficult or impossible to recycle (ceramics, thermosets, and composites).
- Recycling has costs associated with transport and processing (separating, remelting, shredding, etc.), and these must be lower than the cost of using new raw material.
- The highest fractions recycled increase when they require less energy to recycle than to produce (e.g., aluminum).

Designing for Sustainability

Designs can improve sustainably by allowing materials to be easily recovered prior to disposal, reducing unnecessary weight, and considering the material selections from a sustainability standpoint. It is very safe and easy to overdesign something. It is much more difficult to make it just strong enough and thereby improve sustainability. Additionally, allowing a design to be collapsible to reduce shipping volume, considering the efficiency of the electrical components such as motors, and reducing weight all can be pursued in design.

Sometimes sustainability is obvious. Figure 9-1 shows a Karamojong (Uganda) chair perched on the armrest of an American recliner. Note the rope attached to the wooden chair that allows the user to easily carry it from place to place.

FIGURE 9-1. Which chair takes less energy to build, transport, and dispose of?

In your design, consider these questions:

- Where are the raw materials extracted?
- How are they extracted?
- What political, social, and environmental problems do these extractions cause?
- How are the materials transported from the extraction source to the processing location?
- How are they used in the final material?
- How is the final material processed?
- What are the political, social, and environmental concerns that arise from this processing?
- How are the final materials transported to the product manufacturing site?
- How are the final materials used and processed for your product?
- Is your product designed for disassembly (which falls nicely in line with the design for assembly concept that manufacturers love)?
- How is your product disposed?
- Are the materials you are considering recycled or reused? Is this a possibility?

Closing Thoughts

Sustainability sounds good, but how much are we willing to sacrifice for it? That is a personal question that balances many factors, such as economics and comfort, which can't be addressed here. However, when sustainability is kept in mind during design, simple, nonvisible, and no-cost options might arise. For example, a design feature might allow you to separate recyclable materials easily, especially expensive ones. When deciding about material selection, you can consider its sustainability versus cost. You may be surprised that there is no additional cost.

The fundamental goal of this chapter was to encourage understanding of embodied energy and facts that derive from this notion. Heavy things cost more to transport; difficult-to-remove items will not be removed for recycling; and pollution and other externalities are real social costs.

CHAPTER 10

Mechanical Systems

THE PREVIOUS CHAPTERS ESTABLISHED THEORIES THAT PROVIDE A FOUNDATION FOR THE practical application of design and invention. This theory provides deep roots for the diaphanous wash of creativity. Let's now move into some applied aspects of engineering: the mutable nuts and bolts of mechanical engineering practice.

Great inventions can arise from putting together existing products. The clever assembly of currently available items can result in systems that solve important problems. Mechanical systems are usually thought of as an assembly of mechanisms, like a watch. But a system can also include assemblies that move mechanical power (e.g., a car) or energy (e.g., a refrigerator). Typically, these systems are controlled by electronics such as programmable logic controllers (PLCs) or customized microcontrollers. However, mechanical systems can also be controlled by mechanical linkages, such as hydraulic and pneumatic (compressed air) systems.

Products exist that will convert electrical, mechanical, hydraulic, and pneumatic power into linear or rotary motion. These motions can be made with great precision and great force. They can be controlled remotely, such as a radio control airplane, or directly as with a bulldozer. A variety of materials can be moved with precise control, such as cooling air pushed by a fan or thick slurry pumped using a progressive cavity pump. In many ways, systems designs are the most dramatic designs—rocket ships, cars, boats, airplanes; these all fire up the imagination more than something like a staple remover.

Many wonderful inventions have been developed in the world of engineering and they lay at your disposal. Even more helpful, many of the design details are extremely well understood, so you don't need to design gears, bearings, and seals—instead, you can simply specify an existing package. The manufacturers have very practical "cookbook" guidance that allows you to specify every dimension, tolerance, surface finish, and lubricant related to their product. This information is often embedded in CAD programs or available online. As you are reading this, many people serving their niche product line are completely focused on making their product effective, reliable, and cheap. As a designer and inventor, you get to draw on this sophisticated insight and drop these marvelous devices into your product or system. You have access to many great products and you should consider them as if they were parts of a Lego kit that allow you to make things go fast, do complicated mechanics, or operate in extreme environments.

Always remember these systems are dangerous and need to be guarded to prevent anything from being grabbed and ripped apart by these squeezing, scrunching, grabbing mechanical wonders.

Pumps, Compressors, and Fans

Pumps are considered turbomachinery, *turbo* being the Latin word for spin. If a machine adds energy to a fluid, it is called a *pump*; if it depletes the energy, it is called a *turbine*. In fact, pumps can inadvertently be turned into turbines just as their electrical cousins, motors, can be turned into generators.

Although the semantics are interesting, in common usage not all pumps spin and pumps are specifically referred to as machines that add energy only to liquids. If gases are pumped, the pump is called a fan, blower, or compressor. As described in Chapter 5, pressure in English units is referred to as either psia (pounds per square inch, absolute) or psig (pounds per square inch, gauge). Psia is the pressure of a fluid including that produced by the gravitational force on our atmosphere. At sea level, the column of air above us exerts 14.7 psi. Psig is the pressure measured relative to the atmospheric pressure. Therefore, psia is simply psig plus 14.7 psi. There are many other units of pressure, including bar, atmosphere (atm), inches of water (IWC), and the common SI equivalent Pascal (Pa).

There are two categories of pumps and compressors: positive displacement and dynamic. Positive displacement pumps force the fluid to move by sheer mechanical force and move a fixed volume of fluid each cycle. They work by opening a chamber and drawing fluid in. The inlet is then closed and an outlet is opened. Mechanical force pushes the fluid out. Dynamic pumps, on the other hand, work by increasing the velocity of the fluid and "throwing" it out the discharge.

The centrifugal pump is the most common example of a dynamic pump and has the principal advantage of being able to produce high flow rates at low cost and minimal energy consumption. Positive displacement pumps are commonly used in applications when pumping a wide range of viscosities or when high pressures need to be produced at low flow rates.

POSITIVE DISPLACEMENT

In many ways, positive displacement pumps are the easiest to understand. Our body is full of these pumps from our heart to our diaphragm. There are many types of positive displacement pumps. Let's consider three representatives: reciprocating pumps, rotary pumps, and peristaltic pumps. One commonly used positive displacement pump is the reciprocating pump. The reciprocating pump uses a tube with a piston drawn through its center. The piston is pulled up from one end, which creates a vacuum on the opposite end. This vacuum draws up the liquid. In practice, a suction and discharge valve are alternately opened and closed to produce flow.

This is the same principle we use to suck water through a straw. When our diaphragm moves down, it creates a vacuum all the way to our mouths, and in turn, to the straw. As atmospheric pressure works on all the water around the straw, the water is drawn into our mouths. A diaphragm pump is a reciprocating-type pump that works on the principle of a diaphragm moving in and out to draw and dispel liquid. Compressors also use reciprocating cylinders as well as other designs such as scrolls and screws.

Rotary pumps rotate a gear or impeller within a casing to move a fluid, acting much like a rotating paddlewheel propelling a boat. Vane pumps, flexible impeller pumps, and gear pumps are of this type. Peristaltic pumps use a flexible tube that is squeezed by a rotating roller to move the liquid through the tube. It works in the same way fingers can squeeze out an air bubble from under plastic tape.

DYNAMIC

Centrifugal pumps and compressors are the most important member of the dynamic pump family. They are a great design for pumping a lot of fluid at a low cost. The centrifugal pump consists of a paddlewheel-like impeller contained within a casing. It differs from a positive displacement pump in that the impeller does not contact the casing—fluid can flow completely through the pump either forward or backward when the impeller is not turning.

Fluid enters the pump in the center through the casing where it is caught by the rotating impeller and thrown outward into the outer casing. This outer casing or scroll gives the centrifugal pump its distinctive doughnut-like shape and provides an area where the discharged fluid can expand, slow down, and increase pressure.

A dynamic pump will get damaged by cavitation if insufficient inlet pressure is provided. Cavitation is caused when a vacuum is produced. As pressure decreases, the boiling temperature also decreases. This is why water boils at a lower temperature at high altitudes and food requiring boiling must be boiled longer to offset the lower temperature. In the case of cavitation, the pressure developed behind the impeller is so low that the fluid boils at the ambient temperature. During cavitation, the fluid boils and collapses. This action abrades the impeller that produced the low pressure. The damage caused by the boiling can occur at pressures below the boiling point of the liquid due to dissolved gases or lighter fluids boiling out first.

The inlet pressure required to prevent cavitation is available for each pump design. This value is called the net positive-suction head (NPSH). The term "head" is synonymous with pressure and draws an image of the force of water held above an object. The net positive-suction head is dependent on the pressure and velocity of the pump inlet and the vapor pressure of the liquid. Centrifugal pumps commonly require a "flooded suction" where the inlet of the pump is always completely full of fluid but no additional inlet pressure is required. As higher and higher pressures are required, more stages are added to the centrifugal pump. The first stage raises the pressure to drive the second stage, which in turn drives the third stage, and so on. In this manner, high pressures can be achieved with a centrifugal pump with a low inlet pressure while avoiding destructive cavitation.

Electric Motors

Electric motors seem to be in every contraption available. They can be carefully positionable stepper motors or giant, three-phase powerhouses.

Motors are comprised of an armature that spins inside a stator. Typically, wire windings in the armature and stator produce an electromagnetic field that compels the armature to move within the stator. Motors tend to have low torque (or turning force), but they can have high rotational speed and therefore high horsepower. Horsepower is the speed at which the torque can be applied.

The armature turns because it is attracted to and repelled by different portions of the stator. A commutator flips the portions or poles so the rotor is always chasing after the attracting force, thereby rotating. The stator can use coils of wire to produce an electromagnetic force or it can be comprised of permanent magnets.

Power tools typically use DC motors. Compared to AC motors, they have the advantage of easy speed control because the speed is directly related to the supply voltage and the motor torque is controlled by the current. Because cordless drills are cheap and use small DC motors, many prototype machines and linkage mechanisms are initially shaken down using a cordless drill for power.

Electrical Power

Power Factor

The common unit of electrical power is the watt (W), which is equal to the product of voltage and current in amps. However, the apparent power, which determines the heating effect and electrical sizing requirements for an AC system, is commonly called "kVA" (like kilowatts (kW), this value, is the product of kilovolts and amps). The reason for the difference between kVA and kW, is that alternating current systems involving lots of wire that produces a type of resistance called *inductive reactance*. Reactance inhibits alternating current and produces a phase difference between the voltage and current called *power factor*.

The formula for apparent power is:

$kVA = P/F$

where:

P = Power (kW)
F = Power Factor

Therefore, a motor's power dictates how big a device (e.g., fan or impeller) it can drive but kVA determines wire sizing and heat dissipation. This calculation could be made with volts and watts instead of kilovolts and kilowatts, but these phase issues aren't often considered for low-power applications.

Frequency

A motor will operate less efficiently if it runs at an AC frequency different than its rating. Therefore if a 60 Hz motor is run at 50 Hz it will lose power and run hotter. A 50/60 Hz motor rating is a compromise electrical winding that is not as efficient at either 50 or 60 Hz as a single-frequency rate motor.

Variable frequency drives are electronic devices that control motor speed. They take advantage of electric motor design by allowing the input frequency to be varied. Reducing the input frequency slows down the motors and allows dramatic energy savings, especially with large motors.

Transferring Power

Transferring power or changing its direction or speed can be done by a variety of means, and the primary mechanical methods are presented in the following three sections in this chapter. Gears are a great way to transmit or change forces in a small package. If you want to transmit these forces a longer distance, belts and chains are suitable. Finally, if the forces need to be transmitted long distances, hydraulic, pneumatic, or electrical systems are required.

Table 10-1 offers an overview of these power transmitting systems. However, there exists a wide range of variability within each system; for example, pneumatic actuators can be very powerful and electrical actuators can be very fast.

TABLE 10-1. Comparison of precision of movement control, speed, and power between power transmission systems

	Precision	Speed	Power
Mechanical	Good to excellent	Fast	High
Pneumatic	Poor to good	Fast	Good
Hydraulic	Good	Slow	High
Electrical	Excellent	Slow	Poor to good

Pneumatics and Hydraulics

Fluid power offers a larger range of travel than most mechanical or electrical systems and it can be used in dusty or hazardous environments. Great precision and power can be provided by hydraulics, while high speed can be produced by pneumatics. Hydraulic systems allow small hydraulic motors to produce tremendous torque. Pneumatic systems behave much like hydraulic systems even though air is highly compressible. The advantage of pneumatic systems is that they can be open systems with the return air simply vented rather than returned back to the system. However, this venting usually makes a lot of noise. Both systems use a pump, storage system, control, and pressure relief valves. Pneumatics were used extensively for control systems but these fluidic logic systems have been replaced by digital controls.

PNEUMATICS

Pneumatic systems are attractive when something needs to run fast and cool, like a dentist's drill. After using the air to make something rotate or move, it can simply be vented to atmosphere. Typically, a pneumatic system is powered by a compressor that compresses outside air. For large systems, the compressed air is routed into a wet tank that allows entrained water and oil to settle. The air travels from the wet tank to an air dryer that removes nearly all the water from the air system. Air dryers often use desiccants, which are hydroscopic chemicals, to directly absorb the moisture. Other dryer designs use refrigeration units and cyclones to condense the water and separate it from the air. After drying, the compressed air is then routed to a dry tank, which acts like a storage battery by providing a reservoir of compressed air to handle intermittent, high flow requirements. A check valve separates the dry tank from the wet tank. The dry air passes through a filter and usually an oil lubricator before being used by the pneumatic equipment.

HYDRAULICS

Hydraulic systems are great for the application of power. They are similar to pneumatic systems in using a fluid to make something rotate or move. Besides the incredible power capabilities of incompressible hydraulic fluids, the difference with pneumatic systems is that the hydraulic fluid needs to be returned back into the system and can't be vented to the atmosphere as with pneumatic systems.

The principal components in a hydraulic system are pumps, accumulators, controls, reservoirs, lines, hydraulic fluid, heaters, and coolers. The pump takes oil from a reservoir and pushes it through filters, a check valve, and accumulator. Hydraulic pumps produce high pressure at relatively low volume and therefore they are discharged into an accumulator that provides a reservoir of compressed hydraulic fluid. The accumulator serves the same role as the dry tank does for compressed air in pneumatic systems. The accumulator also absorbs the pulsations of the pump.

After the pressurized fluid has been sent to the actuators or motors through some sort of valving, the fluid returns back to the pump. This is called a closed system because all the fluid returns to the starting point. Friction heats up the fluid so coolers are commonly used in reservoirs.

ACTUATORS

Pneumatic and hydraulic systems convert pressure and flow into motion through cylinders, motors, and actuators. Electric systems can also produce rotary or linear motion.

Hydraulic cylinders transform pressure into linear motion and are usually used to operate linkages. Cylinders can be single acting or double acting. A single-acting cylinder allows the hydraulic fluid to enter from only one side of the piston to produce piston rod movement and includes a spring return to the unpressurized state. A double-acting cylinder allows the hydraulic fluid to push the piston in either direction and therefore drives the piston rod in or out. Direction control valves send hydraulic fluid to the desired side of the cylinders. Rotary cylinders provide rotation rather than extension. Pneumatic cylinders have deceleration valves to reduce the venting rate and impacting of the cylinder. Sometimes shock absorbers are used to control the end of the stroke impact.

A variety of valves are available to control cylinder functions, such as a pressure valve that controls a sequence of cylinders so a specified pressure is developed before the next cylinder is operated. This could be used, for example, to ensure a certain gripping force before another cylinder moves the gripped object.

Electrical actuators are very convenient and are available in a wide range of linear and rotary styles. They are especially appealing for highly precise and light gripping application because purpose-built systems are available. They are unsuitable for many hazardous environments, but their use is also limited due to their lack of force and speed in comparison to pneumatic or hydraulic systems. However, they are often coupled with mechanical systems to create high force.

Hydraulic motors have many advantages over electric motors. They are physically small, can run at low speeds, quickly reverse directions, and stall without damage. Rotary actuators are like hydraulic motors except they are usually limited to one revolution or less. These devices produce a slow rotary motion at a very high torque.

Gears

Gears are beautiful and intuitive. They can be mocked up by such means as kids' toys or wooden pegs on plywood disks. Gear sets in metal or plastic can be purchased as assemblies. Machining gears is complex; watching gear shapers carve out the complex geometry of a gear tooth is mesmerizing. Evolutions in nanotechnology are allowing gears and other mechanical systems to be made microscopically small and opening an exciting new world of micromachines. Figure 10-1 shows an example of some tiny gears, large by nanotechnology standards.

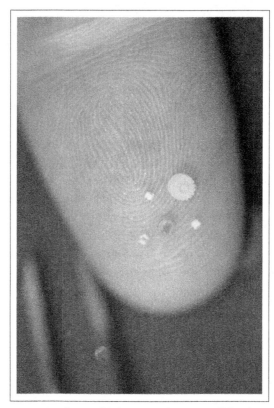

FIGURE 10-1. Even molded gears can be very small (image used with permission from ©Accu-Mold LLC, Ankeny, IA.)

Gear teeth contact each other at a distance called the pitch diameter, which is akin to the outside contact surfaces of a rolling surface, like where a tire meets the road. The gear ratio, which determines the velocity or torque changes, is defined as the ratio of pitch diameters. Figure 10-2 provides an illustration of gear terms.

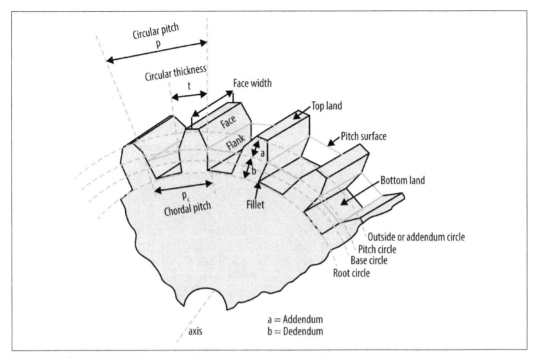

FIGURE 10-2. Gear terminology

Velocity and torque are simple multipliers of the gear ratios. If you have three spur gears spinning, the velocity will be equal to:

$$Vout = Vin \frac{N_1}{N_2} \frac{N_2}{N_3}$$

Or more simply:

$$Vout = Vin \frac{N_1}{N_3}$$

Gears are one of many means to efficiently transfer power. They transfer forces instantaneously and can change rotating or sliding speeds through gear ratios. Rolling wheels can do the same job as gears in terms of changing velocities and torques; however, they tend to slip whereas gear teeth prevent slippage.

To operate efficiently, gear teeth must be opposites (conjugates) of one another such as with the involute shape commonly used in gears. The geometry of the involute tooth must be developed so that the line of action is constant. This line of action works at a certain angle, called a pressure angle, and sets of gears must have the same pressure angle in order to run smoothly and efficiently.

For gear teeth to mate properly, they need to be the same size. That is, they must have the same diametral pitch (or module), which is the number of teeth per length of gear. Specifically, this is equal to the number of teeth divided by the pitch diameter.

Figure 10-3 shows some of the variety of gear types.

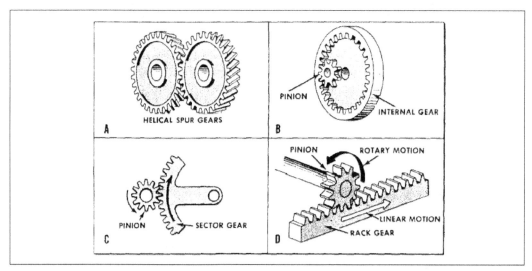

FIGURE 10-3. Basic types of gears

Belts and Chains

Hydraulics and pneumatics allow mechanical energy to be sent far away, while gears allow it to be efficiently sent nearby. Belts and chains are mechanical systems that fall somewhere in between. They operate the same as gears but over larger distances and usually with less speed and torque change. They find special application when a small speed reduction is necessary, such as with electric motors or engines.

While both belts and chains connect two shafts over some distance, belt drives are best for higher-speed, lower-torque applications. Chain drives are best for higher-torque applications. Sometimes moving power over a distance is the only requirement but sometimes these designs are specifically intended to make a change in speed or torque. Gears are better systems when high-speed changes are required, and direct shaft drives are best when you simply want to move power from one place to another.

Pulley systems can operate at higher speeds than chain drives. Typically, chain drives are limited to 1000 ft/min (5 m/s), while belt drives can operate as high as 7000 ft/min (35 m/s).

For both belt and chain drives, at least 120° of the smallest sheave or sprocket should be engaged and they must be parallel and aligned.

BELTS

Belts can be nearly as strong as chains and can run at higher speeds with less noise. Some belts such as the cog type operate very much like chains, while the common V-belt relies on friction between the belt and the pulley. The pulley is usually called a sheave (pronounced "shiv"). While a V belt has contact with a sheave over most of its sides, the pitch diameter is a line of continuous contact, slightly smaller than the sheave's outside diameter, that is used in speed reduction calculations. The speed ratio between the driven and driving sheave is inversely proportional to the ratio of the pitch diameters.

The tension on the belt is a critical but well-understood requirement. High tension causes the belt to burn up and puts load on the shafts, while a low tension allows the belt to slip in the sheave and cause all sorts of trouble. Belt tension is usually controlled with an idler or by allowing adjustment of sheave center to center distances. The typical center-to-center distance of the sheaves is between one and three pitch diameters.

Figure 10-4 shows a Hughes helicopter rotor drive belts. When an idler pulley is tensioned, the engine drives these belts that drive the main and tail rotors.

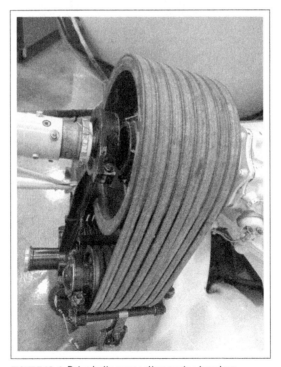

FIGURE 10-4. Drive belts connecting engine to rotors

CHAINS

Chains are great for low-speed/high-torque applications as seen on one of the greatest inventions of all time, the bicycle. As with belts, many chain styles exist but the most common is the roller chain drive with pinned linkages. The pins engage the tooth of a sprocket and transmit the power through the linkages. Because the chain is segmented by pins, the pin spacing becomes important in specifying sprockets. The pin spacing is called pitch and is related to a chain number.

The typical center distances for chains should be between 30 and 50 times the pitch of the chain. Generally, the speed ratio should not exceed seven. Additional stages should be added if you need more reduction.

As with gears and pulleys, the pitch diameter of chains is used to calculate velocity and torque change.

Figure 10-5 shows an engine timing chain.

FIGURE 10-5. Chains are capable of handling high torques

Cable Drive

Cables encased in a tubing can be used to transmit power. The advantage of these systems is that they are relatively inexpensive and can route power through nasty twists and turns. These operate like a bicycle brake cable except they can rotate. These drives are used for low-power and low-speed applications. Cable drives are best used when continuous operation is not required.

Brakes, Clutches and Shaft Drives

Controlling power and aligning drives to shafts is important in designing machinery. Mechanical brakes and clutches have been developed to a fine art. Technological advances are allowing the use of electrical and special fluid properties to intrude into this mechanical realm. Shaft drives keep the output of devices capable of transmitting their power efficiently, durably, and quietly.

BRAKES AND CLUTCHES

Brakes control speed and stop movement while clutches disconnect a power source, such as a motor, from what it is driving. Brakes are excellent examples of the first law of thermodynamics because they turn mechanical work to heat.

Brakes and clutches typically rely on a combination of friction and clamping force, so both a higher-friction material or more clamping force will produce more braking or force transmission. Many mechanisms achieve this, such as the disk brakes on cars, as shown in Figure 10-6. Disk brakes on cars work by pressurized hydraulic fluid pushing brake pads so they drag across a disk. Removing the heat from brakes and clutches requires an understanding of how they are used in their application and lots of experimentation.

FIGURE 10-6. Drum brake

While most clutches use friction to engage one part with another, as shown in Figure 10-7, there are some interesting derivations from the traditional friction surface. For example, magnetorheological clutches use a fluid whose viscosity can be varied, which changes the amount of torque transferred, by an external magnetic force. Eddy current drives rely on the eddy currents induced by moving a conducting disc through a magnetic field to produce braking. Another unique clutch is the single revolution clutch that uses a stop to turn something exactly one revolution.

FIGURE 10-7. Clutch plate is the wearing part of a clutch assembly

Sprag, roller, and cam clutches use their unique geometry to create a wedge between the driving and driven shaft and thereby transmit power between them. But when the input shaft rotates in the opposite direction, the wedges (sprags, rollers, or cams) are no longer wedged and therefore no force is transitted through the clutch.

SHAFT DRIVES

Connecting shafts directly does not provide a speed or torque change, but it is the most efficient system for transfering power, as, for example, between an electric motor and pump or a low-speed engine and propeller. To allow shaft drives to work well a drive is installed inline with the shaft. Drives accommodate misalignment, torsional vibration, and impact loading. One of the most common is the Lovejoy coupling, which includes a rubber (OK, elastomeric) star-shaped insert through which all the power is transmitted.

Closing Thoughts

Mechanical systems may sound quaint in this age of information technology. However, we are largely mechanical. Also, we like mechanical things. We are surrounded by mechanical things and they have a long history of being labor-saving and fun-generating.

This overview gave a flavor for how power is transferred and controlled whether by direct mechanical means, hydraulically, pneumatically, or electrically. Mechatronics, the integration of mechanical, electrical, and information technology, is dynamic and will reveal many amazing developments. Elaborate mechanical governors and pneumatic controls are waning. However, hydraulic power and mechanical drives such as gears and belts will probably enjoy a long future, even in the era of the brilliant machine.

CHAPTER 11

Mechanical Components

THIS CHAPTER PRESENTS THE HEROES OF MANY DESIGNS. THE MANY PARTS AND PIECES NECessary to make a product operate can't be covered here—that is why we have the Internet. However, we should consider the common features in many designs, specifically bearings, gaskets, seals, fasteners, and bonds. While excellent design guides are available for these, they deserve a grand introduction here. Bearings and seals allow things to go very fast, very smoothly, and very reliably. Tiny wisps of metal and goopy adhesives are used to fasten materials and these are often where people intersect with design maintenance, from changing batteries to making upgrades.

Surface coatings are another category of design heroes. Coatings can do magical things in providing appeal, protection, lubricity, heat resistance, electrical conductance, and other vital performance attributes. Good designs are often the orchestration of many small things. Rarely are products some simple, elegant design—rather, they are developed by fussing over the minutiae of hardware specifications, coating, lubrication, part clearances, and dimensional tolerances.

Bearings

Bearings allow things to move easily and are critically important to rotating machinery. For simple, low-speed applications, smooth bushings are used. Bushings don't have rolling elements like balls and rollers; they rely on a very smooth surface finish such as Teflon or internal lubricity such as provided by graphite, molybdenum disulfide, or PTFE (polytetrafluoroethylene). For high-speed applications, a pressurized lubrication system is required for fluid film bushings that require a thin film of oil to lubricate them.

Other important bearing categories are ball, needle, and thrust. Different categories of bearings are suited for different speeds and different environments. The grease used in bearings can be customized for food usage or be delightfully impervious to solvents like fluorinated greases. They can be sealed so they are not contaminated by dirt and debris. An important category when working with prototypes are pillow block bearings (Figure 11-1). This mechanical system includes the bearing and the support housing as one unit. The housing can simply be attached to a frame with bolts and be ready to get to work.

FIGURE 11-1. Pillow block bearing

While bearings last a long time, as they start to wear and approach complete failure, they will make a whirring or grinding sound. Other signs and symptoms of bearing problems are shown in Table 11-1.

TABLE 11-1. Bearing diagnostics

Sign	Problem
Vibration	Loose parts, unbalanced shaft, bent shaft, worn bearing
Hot	No lubrication, too much lubrication, high load or speed
High pitch noise	Misalignment
Low pitch noise	Bearing worn out
Intermittent noise	Loose parts, dirt in bearing, shaft to bearing gap
Excessive wear on shaft or bearing bore	Wrong shaft to bearing bore design, bent shaft, unbalanced shaft

Dimensional requirements are straightforward and manufacturers have helpful application guides that walk you through the process. Application guides will indicate the housing design requirements, such as concentricity, amount of press fit required, and approaches for holding things in place. CAD programs usually have embedded application guides in which you can insert the correct component and support hardware, with dimensions and tolerances, into your drawing.

Figure 11-2 shows a variety of bearing types.

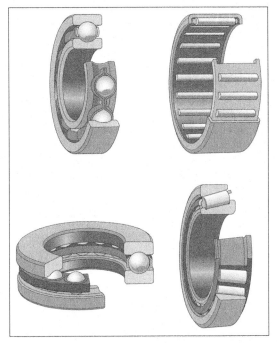

FIGURE 11-2. Ball, needle, tapered roller, and thrust bearings

Gaskets and Seals

Gaskets create a static seal between two parts and can be divided into soft, semi-metallic and metallic categories. Soft gaskets include materials such as elastomers, compressed fiber material, graphite, PTFE (polytetrafluoroethylene), and cork. Semi-metallic gaskets are a composite of metals and nonmetals that allow the wear resistance and strength of metals to be effectively coupled to the flexibility of compressible materials such as seen in spiral wound gaskets. Metal gaskets handle high temperature and are most commonly seen as engine piston rings.

Cork gaskets are pretty much lost to history and often changing the cork valve cover gasket in an old engine was our first foray into gaskets. Modern elastomeric gaskets are made out of materials such as butyl rubber, neoprene, ethylene propylene diene (EPDM), nitrile, silicone, fluoroelastomers, chlorosulfonated polyethylene, and styrene butadiene. Fibrous materials include cellulose, glass, mineral wool, aramids, and carbon fiber. Gasket material selection is motivated by the materials you are trying to control and the environmental conditions with which the design needs to contend.

Shaft seals keep fluids isolated in rotating machinery. They also keep fluids out of an assembly. They can be as simple as O-rings or as complex as mechanical seals. Seals usually use elastomers that slide along the rotating shaft and require a specific squeeze and minimum shaft wobbliness (runout) to produce a good seal. Lip seals (or radial shaft seals) are the most common seal for small, high-speed shafts. This seal is comprised of a lip of elastomer backed by a compression spring. The lip contacts the rotating shaft and the spring puts a gentle force behind it to keep in contact with the shaft. Lip seals can have multiple lips for better sealing of low-speed shafts. Figure 11-3 shows lip seals and typical installation.

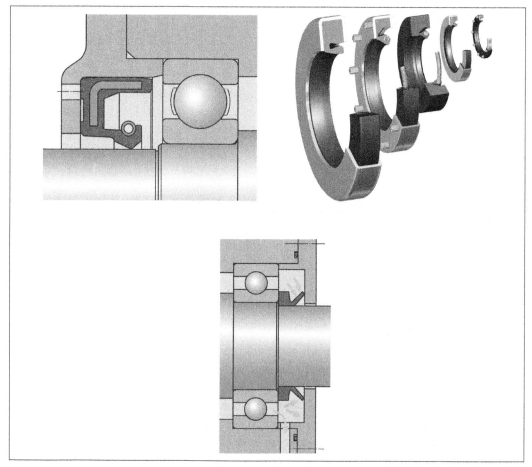

FIGURE 11-3. Common lip seals (or radial shaft seals) and typical installation

The control of squeeze and seal movement dictate the dimensions and tolerances for interfaced parts so that they are the most tightly controlled in any rotating machinery. O-rings can produce an inexpensive shaft seal for slow-speed applications. When parts are not moving against each other, the O-ring thickness can be squeezed down to 40% of its original size. However, when an O-ring is used as a shaft seal, the squeeze must be only about 10% of its original size.

Labyrinth seals are a type of mechanical seal that uses complex geometry to produce a seal without contacting surfaces. This seal is helpful in high-temperature, high-speed applications or when chemical compatibility is a problem.

Fasteners and Bonds

Fastening things together is important and the means by which to do so are incredibly and beautifully varied. Parts and pieces of a design can be held together permanently by welds, adhesives, and interference fits. Parts can be fastened to other things by removable fasteners such as screws, rivets, clips, anchors, pins, retaining rings, staples, buckles, and cable ties. A design's performance, maintenance, and recyclability are all affected by the type of fastening. Welds, adhesives, and press fits can't easily be separated, while screws can be easily removed. Fasteners have a variety of applications; for example, clips allow shafts to rotate at very high speeds while still holding everything in place. Pins allow things to rotate and rivets allow one-sided attachment. Like bearings and seals, there exists a wealth of step-by-step guidelines for creating the correct dimensions and tolerances to allow fasteners to work reliably.

Many new technologies are being applied to fasteners and bonding. Surface-mounted hardware for electronics is fed on a tape and soldered in place to avoid damaging finished circuit boards with clumsy fastener assembly. Long lines ("stick") of fasteners attached tip to head can quickly insert small fasteners that shear off from the following fastener when a specified torque is achieved. Self-healing polymers are being researched that can spontaneously and independently repair themselves.

THREADED FASTENERS

Threaded fasteners describe parts employing a thread to develop clamping force. Screws and bolts are the most commonly used threaded fasteners. The difference between a screw and a bolt is that a bolt is tightened with a nut, whereas a screw is engaged into a tapped hole and is tightened by torqueing its head. Machine screws (which are often used as bolts) are small screws, whereas cap screws are large screws. Bolts are fastened by using a washer, lock washer, and nut or simply a washer and a locking-style nut (e.g., Nylok). Both screws and bolts benefit from thread lubrication not only for corrosion resistance but also to prevent any loosening of their clamping force. Fasteners have to be firmly torqued to 90% of their load capability to ensure their reliability. These values are shown in Table 11-2.

TABLE 11-2. Minimum torque values, foot-pounds

SAE grade	Bolt diameter, inches					
	1/4	3/8	1/2	5/8	3/4	1
2	6	20	47	96	155	310
3	9	30	69	145	234	551
5	10	33	78	154	257	587
8	14	47	119	230	380	700

Metric grade	Bolt diameter, mm								
	6	8	10	12	14	16	18	22	24
4.6	3 (4)	8 (12)	17 (23)	32 (43)	52 (70)	77 (100)	110 (150)	190 (260)	240 (320)
8.8	6 (8)	14 (19)	29 (39)	53 (72)	85 (110)	120 (160)	170 (230)	320 (430)	410 (550)

Torx and square drive heads are rapidly replacing Phillips-head screws. They provide better grip and less propensity to strip. Self-drilling screws make driving wood screws faster and easier, good when you have all hands occupied and are in rapid build mode. Lag bolts are getting replaced by structural screws, even though they look less hefty, because of their high strength and corrosion resistance. Figure 11-4 shows a collection of various fastener heads. Figure 11-5 shows some of the variety of fasteners.

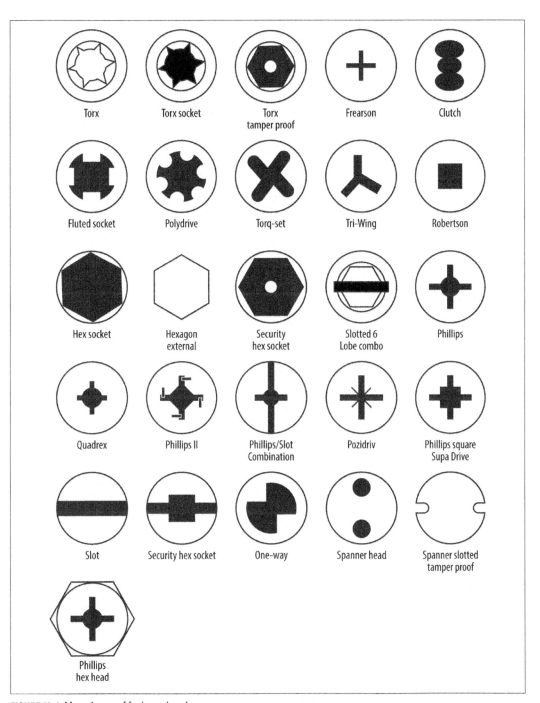

FIGURE 11-4. Many types of fastener heads

FIGURE 11-5. Fasteners come in all shapes and sizes

Even fasteners have been assailed by information technology. While they have long been common subjects for various nondestructive evaluation schemes such as radiography they are now actively evaluating themselves. "Data" bolts, for example, contain a radio frequency identification (RFID) transmitter that is coupled to a small memory circuit and antenna. This system provides information on the bolt's actions during manufacturing that allows them to be traced and improves quality control.

RIVETS

Pop rivets are used for many applications, especially for joining sheet metal or trim to a frame. The advantage of rivets is that they can be quickly installed from one side, like the whole category of blind fasteners. The joint is produced by drilling holes through the materials to be joined and then inserting a rivet. A pin that sticks through one side is pulled with a rivet tool. The pin deforms the back buck-tail. The riveted materials are squeezed between the deformed buck-tail and outward-facing head. Rivets formerly were used for structural steel applications with the rivet being heated and hammered. Rivets are easy to remove by drilling a hole through them and hammering them out of their hole.

SHAFT AND TERMINAL FASTENERS

Keys, pins, clips, and cotter pins are other categories of fasteners commonly used on rotating objects such as gears and pulleys. There are a variety of clips in service. Most of them are spring steel and are intended to engage a shaft slot to hold the shaft in position.

BONDED JOINTS

A bonded joint is made by using a separate material to attach two objects, as is done with brazing, soldering, and adhesion. Some materials are produced by bonds such as plywood and sandwich composite construction, which uses resin adhesion to bond a core and composite laminate. Adhesives were covered extensively in the section "Chacteristics of Adhesives" on page 80. Adhesives can connect almost anything from artificial implants on bones to windshields on cars. The performance of adhesives is usually dependent on surface preparation, temperatures, and humidity. Bonding materials can be very dangerous and the fumes can range from the noxious (such as those emitted by Bondo) to the annoying (such as those produced by heated solder flux).

WELDS

Welding is the process of joining two metal objects by melting the material and fusing them together. The parts to be welded are rigidly held by clamps. It is common for parts to move as they are heated and cooled. An experienced welder will shim parts in anticipation of their movement. It is also common practice to tack weld (a very small, temporary weld) an object to develop the general shape. After the part is in its final shape, continuous welds are made to produce the final object. Residual stresses can linger in these welded joints after the clamps are removed; therefore, welds are often heat treated or hammered (peening) to remove these stresses. Large panels or repetitive welding processes are usually done by robotic or automated welders.

Both material strength and fatigue life in the weld material are reduced by the welding process. This is due to metallurgical changes developed by the high welding temperatures, the shape of the weld, and voids. Generally, the maximum stress allowed in the weld is between 40%–66% of the yield stress of the filler metal, depending on the type of loading. Fatigue strength also is reduced in the weld. The fatigue strength is reduced from 15% for a butt weld to 60% at the end of certain fillet welds.

The two most common welded joints are fillet or butt welds. These welds lay a bead of filler material either at the apex of a joint (in the case of a fillet weld), or at the ends of two materials (in the case of the butt weld). Resistance welding, such as spot welding, is the cheapest welding technique. The weld is made simply by passing an electrical current through the parts to be welded. This current melts the material and produces the weld; this technique eliminates the need for filler rods and fluxes. However, spot welding can only be used on thin sheets (although adhesives are increasingly replacing this technique). The heat-affected zone around these welds is difficult to control and they are not good in tension. A common failure mechanism occurs in spot welds when the hardened weld area tears out under tension.

Figure 11-6 provides a summary of weld types.

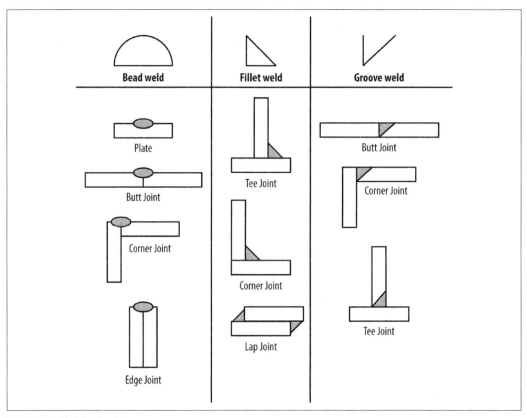

FIGURE 11-6. Variety of weld types

INTERFERENCE FITS

Interference fits can be called force, press, friction, or shrink fits. They are commonly used to bond materials such as shafts to rotors and the like. They are created by forcibly driving one material into another or by heating up and expanding one material and allowing it to shrink around another material. Figure 11-7 shows the range of shaft fits.

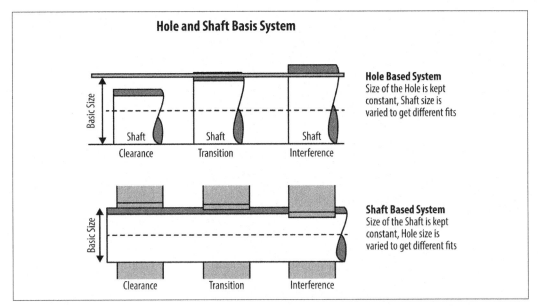

FIGURE 11-7. Types of shaft fits

Springs

Springs can offer quiet heroics in a design. They make things move, stop, and rotate. They balance forces, create clutches, and promote safety. Typical spring types are compression, tension, and torsion. Less common are varieties of spiral, volute, disk, leaf, constant force (like a clock spring or hose clip that gradually releases force), and Belleville spring. While many design guides are available, it is helpful to have a wide variety of springs available when prototyping. There can be a bit of experimentation required to find the ideal spring. Figure 11-8 shows three common types of springs.

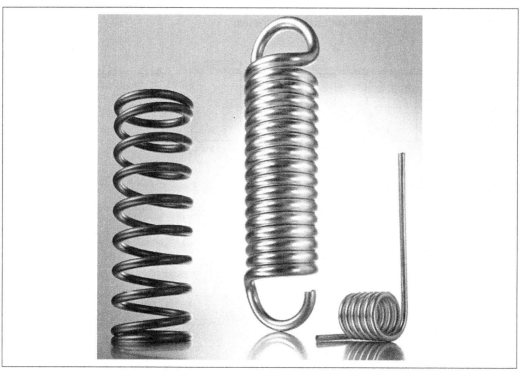

FIGURE 11-8. Commonly used spring varieties: compression, tension, and torsion

Closing Thoughts

Wood, plastic, and metal can be combined to make wonderful things, but mechanical components let your design actually stay together, rotate, and generally function as you want them to. Fortunately, there are hundreds of mechanical components with new innovations coming along all the time. This section highlighted some of the common ones. It did not intend to cover everything available or how to use them. This is changing too quickly for a book to try to capture.

However, the material components you most often use might be the ones with which your are familiar. Because you are making thousands of decisions during the process of design, you can't optimize everything. For example, the design for high-speed rotating machinery is well established from ball bearings to bleed air backed mechanical seals. If you deviate too much from these norms, you are in the realm of research, which is vital, however you increase your chances of failure tremendously.

Epilogue

On the Conceptual Edge

As a child, you watched a leaf fall slowly from a tree. It fell easily along a route of its own design. You stuck your finger in a stream and watched the water rise up and around your finger. You felt a power over the water and marveled at the beautifully contorted flow. You wished to understand the things around you, discover wonders, and create marvels. Discovery comes from wonder, and the topics you learned in this book will be a small part of your design and inventing tool kit. You didn't learn all this stuff just to know it, you learned it so you can use it. You make thousands of decisions when you create something. The engineering background gently offered here will help infuse your design process with some boundaries. You hate boundaries? That's OK, they are not offended.

There are so many ambiguous elements that go into creation that some structure is helpful. The structure of engineering principles doesn't constrain you, it just lets you drive your creative enthusiasm into different directions. So as one channel closes because of some physical reality, you ricochet off the channel and blast through a twisty bend to discover something else. And if you do close a channel because of some "rule of physics" problem, you are closing it with a tired Velcro flap, not a dead-bolted slab of steel.

There is beauty in drawing deeply upon the mind's ability to connect ideas, to steal back the time from forces which wish to squeeze entertainment and communication between every heartbeat. Pausing to think is the opposite of the keyword searching approach to acquiring knowledge, with its uninvited AI curating. These pauses can provide insight, empathy, curiosity, and ambition. Reading this book, or any such book, is a challenge. It required focus and processing. It competes for your time in an era when technology is always your partner—when smartphone-coupled conversation encourages superficial topics that easily allows us to check in and out.

This book is intended for those working at the conceptual edge of the design process. It is for the inventor who sees a need and ingeniously creates in the void. It is for the industrial designer who brings user-centric focus and connoisseurship into design. Unlike intrinsic material properties and anthropometric data, the boundaries for those at the conceptual edge should always be pushed back. Those who work with concepts operate across disciplines and constantly challenge the status quo.

The Process

Concept development is just part of the design process. The process may start with a crafted marketing goal or an individual's great idea. The design brief is then pursued in concept development, which typically involves a cycle of developing empathy for the user, making sketches, and testing models. Concept development is followed by traditional engineering, where a design is analyzed and detailed, performance standards achieved, and quality assurance procedures developed. The design is then prototyped and tested before releasing to manufacturing. While this is a simple string of words, the reality is very complex. Not only is the process complex but it becomes increasingly difficult to make changes to a design the further it progresses in the process. Engineers churn a concept to develop a polished jewel. Manufacturing often makes expensive capital investments, so changing something like a multi-cavity injection mold can be a costly mistake. So too can be reworking advertisements or modifying performance standards. All of these processes and expenses are based on a concept, which comes from a time when all ideas are options and the "great idea" is waiting around the corner. The paper is cheap and the time gladly spent.

This book looked at a variety of nontechnical issues, but focused on material mechanics, fluid mechanics, heat transfer, and thermodynamics. These disciplines are the cornerstone of the mechanical sciences. Mechanical science works in the company of the other cornerstones of mechanistic design: the biological, electrical, and information sciences. Design, in the broad sense of both mechanistic and nonmechanistic pursuits, relies on many disciplines. The physical sciences, life sciences, math, anthropology, sociology, psychology, and economics can all be tapped for methods that improve design. Knowledge derived from these disciplines improve design and therefore we must borrow from them with alacrity and continually explore.

The Next Page

We are all looking for the elegant, simple design. One that has never been thought of. One that reflects the power of our mind infused with lightning strikes of imagination. The other side of this pursuit is to embrace complicated problems that require complicated solutions. Complex problems are solved one step at a time in a weird, circular path that keeps out the amateurs.

Please take your great ideas—the ones that rise above the noise of the culture and whisper restlessly looking for a home—and draw them. Now build them, even as an ugly, contorted mass of cardboard, clay, and wood. Make them come to life. The process is beautiful. This study of engineering and design was not about how to design—it was about your mind. It was giving your mind some foundational concepts. Concepts that do not surf on trends, but ride on immutable laws of physics that offer a foundation for the mechanical side of wonder-making. Driving knowledge into the tacit realm, involves harmonizing "book learning" with experience and forcing your mind to stitch everything together. You want to walk away with more than facts; you want to walk away with a feel for how mechanical designs behave.

Now go. Design and build. Have fun and keep exploring.

Index

A
abduction, 147
abrasive wear, 74
absolute pressure, 90
absolute zero, 90, 95
absorption, 125
acceleration of objects by loads, 36
acoustic resonance, 61
action and reaction, Newton's third law regarding, 37
actuators, 165
adduction, 147
adhesives, 80-82
aestheticism, 14
aesthetics, 3, 9, 13-19
affordances, 26
aircraft carrier design, 14
aircraft design, 2, 135
aluminum, 70-71, 73, 86
analogous colors, 28
anisotropic materials, 43, 67
    (see also composite materials)
anode, 82
anthropometric data, 140-144
aramid fiber, 80
Armstrong, Neil, 7
art, vocabulary for, 26
    (see also aesthetics; creativity)
artificial neural networks, 149-150
Ask, Thomas, ix
atmospheric pressure, 90

B
balance, 27
baseline temperature, 90
bearings, 175-177
beauty, 15-16
belt drives, 168-169
benchmarks, 50
Bernoulli equation, 94
Bingham plastic, 113
biological corrosion, 85
biomimicry, 148
blood, 113
bonded joints, 183
bonds, 179-184
boundary layer, 114, 119
boundary layer separation, 118
bounded rationality, 3, 11, 26
brakes, 171
brand management, 19-20
brass, 71, 73, 87
breakage
    deforming prior to (see deformable materials)
    forces causing, 34
    strength determining point of, 36
brinelling, 59
brittle fracture, 58
brittle materials, 39, 49
bronze, 72-72, 87
buckling, 47, 52, 59
building, design by, 4-5
bulk modulus elasticity, 109
bushings, 175

C

CA (cyanoacrylate adhesive), 81
cable drives, 170
carbon, 68
carbon fiber, 80
carpal tunnel syndrome, 144, 146-147
cast iron, 86
cathode, 82
cavitation, 59, 161
centrifugal pumps, 161
ceramic matrix composite (CMC), 79
ceramics, 75
cetane, 105
chain drives, 168-170
cloud point of fuel, 105
CLTD (cooling load temperature differential), 132
clutches, 171-173
CMC (ceramic matrix composite), 79
collaboration, 30
color, 28, 124
combustion, 103-105
complementary colors, 29
composite materials, 43, 78-80
compression, 34
    buckling caused by, 52
    thermal expansion causing, 55
compressors, 160-162
conduction, 122, 128, 136
constraints, 27
contact information for this book, xiii
context, 26
convection, 122-123, 129, 136
convective heat transfer coefficient, 123
cool colors, 29
cooling load factor, 131, 137
cooling load temperature differential (CLTD), 132
copper alloys, 87
coronal (frontal) plane, 147
corrosion, 82-87
couple (see moment (torque))
creativity, 29-31
creep, 51
crevice corrosion, 85
cubital tunnel syndrome, 147
cultural constraints, 27
cultural heritage, 27
culture (see ethnography; material culture; tradition)
cyanoacrylate adhesive (CA), 81

D

deformable materials, 39-43
density, 66, 99, 109
design, xii
    (see also mechanical design)
    aesthetics, role in, 3, 9
    by building, 4-5
    compared to fine arts, 19
    design thinking method, 5-6
    ethnography, role in, 3, 6
    intuition, role in, 1-3
    multiple inputs of, 25-26
    philosophical foundations of, 10-12
    process for, 3-6, 8-10
    starting, 8-10
    virtual world affecting, 7-8
    vocabulary for, 26-29
dew point, 107
dialectic, 4
digital world (see virtual world)
dilatant liquids, 113
dimensional stability, 66
directional movement, 28
disposal of products, 154-155
drag, 116-120
drag coefficient (Cd), 117
ductile materials, 39, 49
ductile rupture, 57
dynamic pressure, 90
dynamic pumps, 161

E

80/20 rule, 149
Einstein, Albert, 10
elastic deformation, 57
elastic limit (see yield strength)
elastic materials, 40
elastic region, 41
elastomers, 75
electric motors, 162-163
electrical power, 162
electromagnetic radiation (see radiation)
embodied energy, 152-154
emission, 126-127
energy
    conservation of, 91-94
    direction of, 95
    embodied in materials or products, 152-154
    minimum energy condition of a material, 95
    movement of (see thermodynamics)

used by products, 154
engineering, mechanical (see mechanical engineering)
entropy, law of, 95
epoxy, 81
equilibrium, 37
ergonomics, 139-148
    anthropometric data for, 140-144
    handle design, 140
    kinesiology for, 144-148
ethnography, 3, 6, 24-25
evaporation, 95-103, 106, 132
exemplar, 27
expansion, 100
experiential learning, 4
expert systems, 149-150
extension, 147

F

failure modes, 49-60
fans, 160
fasteners, 179-184
fatigue, 49, 58
fatigue failure, 50
fatigue strength, 50
fatigue zone, 50
Feyerabend, Paul, 11
fiber materials, 79-80
fiber orientation, 43
fiber reinforced plastic (FRP), 43, 79
fiberglass, 80
fibers, 79
Fibonacci spiral, 10
film conductance, 123
fishing vessel design, 3, 23
Fitt's law, 149
flammability, 66
flash point of fuel, 105
flexion, 147
flexure, 35, 43
fluid dynamics, 112-120
fluid statics, 110
fluids, 109-120
    boundary layer of, 114
    convection using, 122-123
    dimensionless parameters of, 115-116
    drag, 116-120
    hydraulic systems using, 111-112, 164
    Pascal's law of, 110
    pumps and turbines for, 160-162

    rheology, 112
    viscosity, 109
foams, 76
forces
    effects on solids, 34-43
    free body diagram of, 38
    Newton's second law regarding, 37
    reaction to (see moment (torque))
    types of, 34-36, 54
    units for, 40
form, 27
Fourier's law, 122
free body diagram, 38
frequencies of sound, 61
frequency, 137
fretting fatigue, 51
frontal (coronal) plane, 147
FRP (fiber reinforced plastic), 43
fuels, 105

G

galling, 59
Galvani, Luigi, 82
galvanic corrosion, 82
gas phase, 95-103
gases
    compressors and fans for, 160-162
    convection using, 122-123
    infiltration using, 133
    mass transfer using, 132
    pneumatic systems using, 111-112, 164
gaskets, 177-179
gauge pressure, 90
gears, 165-168
gel point of fuel, 105
glass transition temperature, 66, 75
glossary
    of art and design, 26-29
    of kinesiology, 147
golden ratio, 9
golf balls, surface of, 119
gray body, 126
Grof, Stansilav, 11
group identity, 20-21, 29, 31

H

handle design, 140
hands, anatomy of, 147
hardness, 65-66

harmony, 28
heat, 121
heat transfer, 121-138
    conduction, 122, 128
    convection, 122-123, 129
    internal sources of heat, 135
    mass transfer, 132
    radiation, 123-132
    specific heat, 128
    thermal storage, 134
    types of, 121
Heidegger, Martin, 4
Herbert, George, 9
homogenous materials, 67
Hooke's law, 40
human factors, 139-150
    artificial neural networks, 149-150
    biomimicry, 148
    ergonomics, 139-148
    expert systems, 149-150
    interaction design, 148
    mapping visual cues to functions, 145
humanism, 2
Hume, David, 11
Husserl, Edmund, 3
hydraulic systems, 111-112, 164

I

ideal gas law, 91-91
impact, 58
impact resistance, 66
inertia, 37
infiltration, 132-133
informant, 27
infrared light, 124
instantaneous zone, 50
instrumentalism, 14
interaction design, 148-149
interference fits, 184
intergranular corrosion, 85
Internet (see virtual world)
intuition, role in design, 1-3
isotropic materials, 43, 67
    (see also metals)

J

Joule-Thomson coefficient, 100
Jung, Carl Gustav, 29

K

Kant, Immanuel, 10
kinesiology, 144-148
Kolb, David, 4
Kuhn, Thomas, 10

L

laminar boundary layer, 114
laminates, 79
latent heat, 131
learned helplessness, 27
light, 123-124
liquid phase, 95-103
liquids (see fluids)
loads
    applied to an object (see forces)
    static, 36-39
    types of, 54

M

Mach (Ma) number, 115
magnesium, 72-73
malleable iron, 73
mapping, 27, 145
material culture, 21
materials, 34
    (see also fluids; gases; solids)
    categories of, 67
    classification of, 67
    properties of, 65-66, 124-127
mean radiant temperature (MRT), 131
mechanical components, 175-185
    bearings, 175-177
    bonds, 179-184
    fasteners, 179-184
    gaskets, 177-179
    seals, 177-179
    springs, 185
mechanical design, xii
    (see also fluid dynamics; fluid statics; heat transfer; thermodynamics)
    aesthetics, role in, 3, 9
    by building, 4-5
    design thinking method, 5-6
    ethnography, role in, 3, 6
    human factors in (see human factors)
    intuition, role in, 1-3
    materials used in (see fluids; gases; materials; solids)

multiple inputs of, 25-26
philosophical foundations of, 10-12
process for, 3-6, 8-10
starting, 8-10
sustainability in (see sustainability)
mechanical systems, 159
    actuators, 165
    belt drives, 168-169
    brakes, 171
    cable drives, 170
    chain drives, 168-170
    clutches, 171-173
    compressors and fans, 160-162
    electric motors, 162-163
    fans, 160
    gears, 165-168
    hydraulic systems, 111-112, 164
    pneumatic systems, 111-112, 164
    pumps and turbines, 160-162
    shaft drives, 173
    for transferring power, 163
mechanistic influences, 13, 26
metal matrix composite (MMC), 79
metals, 43, 67-73
    aluminum, 70-71
    brass, 71
    bronze, 72-72
    magnesium, 72-73
    nickel, 72
    stainless steel, 69-70
    steel, 68-69
    steel alloys, 69
    strengths of, 73
    titanium, 73
    zinc, 73
MMC (metal matrix composite), 79
modulus of elasticity, 40-43, 66
moment (torque), 36
moment of inertia, 44-47
motion, Newton's laws of, 37
motors, electric, 162-163
movement of objects by loads, 36
MRT (mean radiant temperature), 131
Munch, Edvard, 30
muscular-skeletal disorders (MSD), 146

N

net positive-suction head (NPSH), 161
Newton, Isaac, 10
Newtonian fluids, 112

Newton's laws of motion, 37
nickel, 72
nickel alloys, 73, 87
noise, 61-62
nontechnical (nonmechanistic) influences, 13, 26
    aesthetics, 13-19
    brand management, 19-20
    creativity, 29-31
    ethnography, 24-25
    material culture, 21
    tradition, 22
    visual stereotypes, 22-23
NPSH (net positive-suction head), 161

O

octane, 105
opaque, 28
orthotropic materials, 67

P

packaging of products, 153
Pareto principle, 149
Pascal, Blaise, 3
Pascal's law, 110
people, 2
    (see also human factors; humanism)
    comfort of, 131, 139
    ergonomics for, 139-148
    heat produced by, 131, 135, 137
    injuries related to design, 146
    movements of, 144-148
    study of (ethnography), 3, 6
peristaltic pumps, 161
phases of molecules, 95-103
philosophical foundations of design, 10-12
physical constraints, 27
pickup truck design, 23
pitting, 85
plastic strain (see yielding)
plastics, 66, 75-76
PMC (see polymer matrix composite)
pneumatic systems, 111-112, 164
Poisson's ratio, 41
polar moment of intertia, 46-47
polymer matrix composite (PMC), 79
polymers, 75
    (see also plastics)
polyurethane, 81
polyvinyl acetate (PVA), 81

Popper, Karl, 10
positive displacement pumps, 160
positivism, 2
pressure, 90, 98-99
Pressure Sensitive Adhesive (PSA), 81
primary colors, 28
pronate, 147
proportion, 27
Protagoras, 3
PSA (Pressure Sensitive Adhesive), 81
pseudoplastic fluids, 113
pumps, 160-162
PVA (polyvinyl acetate), 81
PVC cement, 81

R

radial tunnel syndrome, 147
radiation, 123-132, 137
    combined effects with, 128-132
    heat transfer, 123-127
    types of, 123-124
randomness, law of, 95
rationalism, 2
rationality, bounded, 3
reaction, Newton's third law regarding, 37
reciprocating pumps, 160
recycled fraction, 155
recycling, 152
reflection, 125
reimforced plastics, 79
resins, 79
respondent, 27
Reynolds number (Re), 115
rheology, 112
rhythm, 28
rigidity, 36, 44
Rilke, Rainer Maria, 15
rivets, 182
rotary pumps, 161
running, 145
rust, 83-84

S

sagittal plane, 147
sandwich construction
    corrugated cardboard, 44
    fiber reinforced plastic, 43
Schawlow, Arthur, 30
scientific method, limitations of, 11-12
scissors design, 145
seals, 177-179
seat design, 143
secondary colors, 28
section modulus, 46
semantic constraints, 27
sensible heat, 131, 137
shade, 29
shaft drives, 173
shaft fasteners, 183
shape, 27, 43-47
    relationship to stress concentration, 47-49
    stiffness determined by, 43
shear, 34, 43, 109
ski boot materials, 87
Socrates, 7
solids
    categories of, 67
    ceramics, 75
    composite materials, 78-80
    deformable, behaviors of, 39-43
    failure modes of, 49-60
    fiber materials, 79-80
    foams, 76
    forces affecting, 34-43
    metals, 67-73
    plastics, 75-76
    rigidity of, 36
    shape considerations of, 43-47
    strength of, 36
    surface wear of, 74
    toughness of, 43
    wood, 77-78
sound, 61-62
spalling, 60
specific heat, 128, 138
springs, 185
squeezing (see compression)
stack effect, 133
stainless steel, 69-70, 86
standard deviation, 141
static loads, 36-39
static pressure, 90
steel, 68-69, 73, 86
steel alloys, 69
stiffness, 43
stoichiometric reaction, 103
storyboard, 6
strain, 36, 40-43
strength, 36
    fatigue strength, 50

of metals, 73
of plastics, 76
tensile strength, 50
ultimate strength, 41-42, 66
yield strength, 41-42, 66
stress, 34, 40-43
(see also failure modes; force)
stress concentration, 47-49, 50
stress corrosion cracking, 58, 85
stretching (see creep; strain; tension; thermal relaxation)
supinate, 147
surface coatings, 175
surface fatigue wear, 74
surface wear, 74
sustainability, 151-156
    design considerations, 155-156
    disposal of product, 154-155
    embodied energy in materials or products, 152-154
    energy source of product, 154
    recycling, 152, 155
    use and misuse of products, 154

T

temperature (see heat; thermodynamics)
temperature measurements, 95
tensile strength, 50
tension, 34
terminal fasteners, 183
terminology used in this book, xii
(see also vocabulary)
texture, 27
thermal conductivity, 122
thermal expansion, 54-57
thermal expansion coefficient, 122
thermal load, 122
thermal relaxation, 51
thermal resistance factor, 136
thermal shock, 59
thermal storage, 134
thermodynamic table, 93-94
thermodynamics, 89-107
    baseline temperature, 90
    combustion, 103-105
    first law of, 91-94
    ideal gas law, 91-91
    phase changes, 95-103
    pressure, 90
    second law of, 95, 121

    third law of, 95
    zeroth law of, 95
thermoplastics, 75
thermosets, 75
thickness, 44-47
(see also shape)
thoracic outlet syndrome, 146
threaded fasteners, 179-182
tint, 29
titanium, 73
torque (see moment)
torsion, 34, 46-47
toughness, 43, 66
tradition, 22
translucent, 28
transmission, 124
transparent, 28
transverse plane, 147
trees
    design of, 56
    material from (see wood)
tribochemical wear, 74
turbines, 160-162
turbulent boundary layer, 114, 119
twisting force (see torsion)

U

ultimate strength, 41-42, 66
ultraviolet resistance, 66
units of measure
    for electrical power, 162
    for force, 40
    for stress, 40
    for thermal conductivity, 122

V

value, 28
Van der Waals equation, 101
vibration, 62
virtual world, effects on design, 7-8
viscoelastic materials, 34, 110
viscosity, 105, 109
viscous materials, 34
visible light, 123-124
visual appeal (see aesthetics)
visual stereotypes, 9, 22-23, 26
vocabulary
    of art and design, 26-29
    of kinesiology, 147

W

walking, 144
warm colors, 29
watch design, 14
water absorption, 66
wave drag, 119
wavelength, 137
wear, 59, 74
weather
    dew point, 107
    radiation heat transfer, 123-127
    wind, 133
website resources, xiii

welds, 183
wind, 133
wood, 77-78

Y

yield strength, 41-42, 66
yielding, 41, 57
Young's modulus (see modulus of elasticity)

Z

zinc, 73

About the Author

Thomas Ask is professor of industrial design at the Pennsylvania College of Technology and previously worked in industry as a practicing engineer for nearly 20 years. During his time in industry, he designed dozens of commercialized products and systems. Tom's employment includes senior engineer at Ingersoll Rand, principal of Ask and Associates, and vice president of engineering at Odin Systems. He is the founder of the Society of Inventors and Mad Scientists, as well as a licensed Professional Engineer with a doctorate in industrial design.

Colophon

The animal on the cover of *Engineering for Industrial Designers and Inventors* is a Freshwater butterflyfish (*Pantodon bucholzi*). It is part—and the only species—of the Pantodontidae family and can be found in warm, slow moving bodies of water in western Africa, which is why they are also known as African butterflyfish.

Freshwater butterflyfish are pretty small in length, rarely exceeding five and a half inches. The "butterfly" in their name is derived from their beautiful and useful pectoral fins. These fins are widespread, of a yellowish-green to brown silvery base, and marked with spots and lines of darker shades. The body has similar coloring to the fins, minus the dark markings.

The freshwater butterflyfish has a carniverous diet consisting primarily of insects it feeds on at or above the surface of water. Its mouth is upturned to aid in this type of feeding, and the pectoral fins described previously allow the fish to jump out of the water to hunt and avoid predators. The eyes of this fish are also placed on top of the head for better watch of the water's surface.

Breeding for the freshwater butterflyfish happens fairly quickly. Fertilization of eggs is done internally after a male clutches a female between his fins. The female will release a few hundred eggs over a period of days. These eggs are light in color, getting darker as they get ready to hatch. The eggs typically hatch between three to seven days. In aquariums, it is recommended that the eggs be separated from parents, as the parents will likely eat the eggs before they hatch.

Many of the animals on O'Reilly covers are endangered; all of them are important to the world. To learn more about how you can help, go to animals.oreilly.com.

The cover image is from *Lydekker's Royal Natural History*. The cover fonts are URW Typewriter and Guardian Sans. The text font is Scala Pro; heading fonts are URW Typewriter and Benton Sans.

Get even more for your money.

Join the O'Reilly Community, and register the O'Reilly books you own. It's free, and you'll get:

- $4.99 ebook upgrade offer
- 40% upgrade offer on O'Reilly print books
- Membership discounts on books and events
- Free lifetime updates to ebooks and videos
- Multiple ebook formats, DRM FREE
- Participation in the O'Reilly community
- Newsletters
- Account management
- 100% Satisfaction Guarantee

Signing up is easy:

1. Go to: oreilly.com/go/register
2. Create an O'Reilly login.
3. Provide your address.
4. Register your books.

Note: English-language books only

To order books online:
oreilly.com/store

For questions about products or an order:
orders@oreilly.com

To sign up to get topic-specific email announcements and/or news about upcoming books, conferences, special offers, and new technologies:
elists@oreilly.com

For technical questions about book content:
booktech@oreilly.com

To submit new book proposals to our editors:
proposals@oreilly.com

O'Reilly books are available in multiple DRM-free ebook formats. For more information:
oreilly.com/ebooks

O'REILLY®

Have it your way.

O'Reilly eBooks

- Lifetime access to the book when you buy through oreilly.com
- Provided in up to four, DRM-free file formats, for use on the devices of your choice: PDF, .epub, Kindle-compatible .mobi, and Android .apk
- Fully searchable, with copy-and-paste, and print functionality
- We also alert you when we've updated the files with corrections and additions.

oreilly.com/ebooks/

Safari Books Online

- Access the contents and quickly search over 7000 books on technology, business, and certification guides
- Learn from expert video tutorials, and explore thousands of hours of video on technology and design topics
- Download whole books or chapters in PDF format, at no extra cost, to print or read on the go
- Early access to books as they're being written
- Interact directly with authors of upcoming books
- Save up to 35% on O'Reilly print books

See the complete Safari Library at safaribooksonline.com

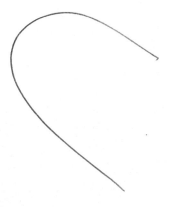

Engineering for Industrial Designers and Inventors

If you have designs for wonderful machines in mind, but aren't sure how to turn your ideas into real, engineered products that can be manufactured, marketed, and used, this book is for you. Engineering professor and veteran maker, Tom Ask, helps you integrate mechanical engineering concepts into your creative design process by presenting them in a rigorous but largely nonmathematical format.

Through mind stories and images, this book provides you with a firm grounding in material mechanics, thermodynamics, fluid dynamics, and heat transfer. Students, product and mechanical designers, and inventive makers will also explore nontechnical topics such as aesthetics, ethnography, and branding that influence product appeal and user preference.

- Learn the importance of designing functional products that also appeal to users in subtle ways
- Explore the role of aesthetics, ethnography, brand management, and material culture in product design
- Dive into traditional mechanical engineering disciplines related to the behavior of solids, liquids, and gases
- Understand the human factors of design, such as ergonomics, kinesiology, anthropometry, and biomimicry
- Get an overview of available mechanical systems and components for creating your product

Thomas Ask is professor of industrial design at the Pennsylvania College of Technology. He previously worked as a practicing engineer for nearly 20 years, designing dozens of commercialized products and systems. Formerly vice president of engineering at Odin Systems and principal of Ask and Associates, he is the founder of the Society of Inventors and Mad Scientists.

> "This book fills a much needed void. Those without a formal engineering and/or math background but who still hold the responsibility to insure that the products that they design actually work will find this book an excellent guide. Engineering concepts are made easy to understand with excellent, real life examples in an undiluted presentation. An invaluable tool for any designer or inventor."
>
> **—Georgene Rada**
> Vice President Global Design, Briggs & Riley Travelware

> "[This] book is really amazing. It is able to transform the engineering jargon into lucid, imaginary visualization."
>
> **—Henry Kang, Ph.D.**
> Head of Marine Technology Laboratory, Universiti Teknologi Malaysia

ENGINEERING / ARCHITECTURE

US $39.99   CAN $45.99
ISBN: 978-1-491-93261-2

Twitter: @oreillymedia
facebook.com/oreilly